Andreas Vesalius

The Reformer of Anatomy

James Moores Ball

Alpha Editions

This edition published in 2024

ISBN : 9789366389615

Design and Setting By
Alpha Editions
www.alphaedis.com
Email - info@alphaedis.com

As per information held with us this book is in Public Domain. This book is a reproduction of an important historical work. Alpha Editions uses the best technology to reproduce historical work in the same manner it was first published to preserve its original nature. Any marks or number seen are left intentionally to preserve its true form.

Contents

PREFACE ....- 1 -
INTRODUCTION ....- 4 -
CHAPTER FIRST Anatomy in Ancient Times......................- 15 -
CHAPTER SECOND Mondino, the Restorer of Anatomy .....- 24 -
CHAPTER THIRD Mondino's Successors.............................- 30 -
CHAPTER FOURTH Vesalius's Early Life ....- 42 -
CHAPTER FIFTH Sojourn in Paris ....- 45 -
CHAPTER SIXTH Vesalius Returns to Louvain....................- 55 -
CHAPTER SEVENTH Professor of Anatomy in Padua.........- 57 -
CHAPTER EIGHTH First Contribution to Anatomy..............- 63 -
CHAPTER NINTH Publication of the Fabrica ....- 65 -
CHAPTER TENTH Publication of the Epitome ....- 73 -
CHAPTER ELEVENTH Contents of the Fabrica...................- 76 -
CHAPTER TWELFTH Contemporary Anatomists ....- 86 -
CHAPTER THIRTEENTH Commentators and Plagiarists ....- 94 -
CHAPTER FOURTEENTH The Court Physician ....- 97 -
CHAPTER FIFTEENTH Pilgrimage and Death....................- 100 -
FOOTNOTES ....- 103 -

PREFACE

In the annals of the medical profession the name of Andreas Vesalius of Brussels holds a place second to none. Every physician has heard of him, yet few know the details of his life, the circumstances under which his labors were carried out, the extent of those labors, or their far-reaching influence upon the progress of anatomy, physiology and surgery. Comparatively few physicians have seen his works; and fewer still have read them. The reformation which he inaugurated in anatomy, and incidentally in other branches of medical science, has left only a dim impress upon the minds of the busy, science-loving physicians of the nineteenth and twentieth centuries. That so little should be known about him is not surprising, since his writings were in Latin and were published prior to the middle of the sixteenth century. His books, which at one time were in the hands of all the scientific physicians of Europe, are now rarely encountered beyond the walls of the great medical libraries of the world. They are among the *incunabula* of the medical literature. That English-speaking physicians know little of Vesalian literature is due to the fact that no extensive biography of the great anatomist has appeared in our language. Most of the Vesalian literature which has been written by English and American authors has been in the form of brief articles for the medical press; these oftentimes have been incorrect and unillustrated. Perhaps the best example of this class is the article by Mr. Henry Morley which appeared originally in *Fraser's Magazine*, in 1853, and later was published in his *Clement Marot and Other Studies*, in 1871. The chief data for Vesalius's biography are to be found in his own writings, in the archives of the Universities in which he taught, and in the controversial literature of the period. Extensive as are these sources they leave much to be desired. A vast mass of Vesalian literature was printed, chiefly in the Latin language, during the seventeenth and eighteenth centuries. Much of it is based on insufficient evidence or on national prejudice. The Germans, the French, the Dutch and the Italians have all taken a turn at it. In modern times the monumental work of Roth, *Andreas Vesalius Bruxellensis*, Berlin, 1892, has served to epitomize this literature and to make clear many points which formerly were not understood. I have taken Roth's book as a basis for this monograph, without using the voluminous references which are found in the work of this thorough historian.

The man who overthrew the authority of Galen; revolutionized the teaching of the structure of the human body; started anatomical, physiological, and surgical investigation in the right channels; first correctly illustrated his dissections; destroyed ancient dogmas, and made many new discoveries—this man, Andreas Vesalius of Brussels, deserves the name which Morley has given him, "the Luther of Anatomy."

At long intervals a bright particular star appears in the intellectual horizon, endowed with genius of such a superlative order as seemingly to comprise within itself the whole domain of an entire science. These men do not belong to any particular epoch in the development of the human mind. They are the eternal symbols of progress, and their history is the history of the science which they profess. Such men were Bacon, Galileo, Descartes, Newton, Lavoisier, and Bichat; and such also was Andreas Vesalius the anatomist. Young, enthusiastic, courageous and diligent, Vesalius dared to contradict the authority of Galen, corrected the anatomical mistakes of thirteen centuries and before his thirtieth year published the most accurate, complete, and best illustrated treatise on anatomy that the world had ever seen. His industry, the success which crowned his efforts, the jealousies which his discoveries aroused in the breasts of his contemporaries, the honors which were conferred upon him by Charles the Fifth and Philip the Second, his pilgrimage to the Holy Land, and his tragic death—these are events which deserve to be chronicled by an abler pen than mine.

The year 1543 marks the date of a revolution which was won, not by force of arms but by the scalpel of an anatomist and the hand of an artist. The whole of human anatomy, as a study involving correct descriptions of the component parts of the body and accurate delineations thereof, may be said to have been founded by Andreas Vesalius and Jan Stephan van Calcar. As light pouring into a prism attracts little notice until it emerges in iridescent hues, so it was with anatomy: after passing through the brain of Vesalius it bore rich fruit which has been gathered by many hands. To turn from the writings of Galen, Mondino, Hundt, Peyligk, Phryesen, and Berengario da Carpi to the beauties of Vesalius's *De Humani Corporis Fabrica* is like passing from darkness into sunlight. To both anatomists and artists this book was a revelation. For more than a century after its appearance the anatomists of Europe did little more than make additions to, and compose commentaries upon the conjoint triumph of Vesalius and van Calcar. For more than two centuries the osteologic and myologic figures of the *Fabrica* formed the basis of all treatises on Art-Anatomy.

JAMES MOORES BALL.

Saint Louis,

MDCCCCX.

I. van Kalker p. I. Tröyen s.
ANDREAS VESALIUS
(From an old copperplate engraving)

INTRODUCTION

The intelligent student of medical history has at his command an unfailing source of pleasure. To learn the successive steps by which Medicine has advanced from a priest-ridden and secret art practiced with mysterious rites in the Greek temples, passing through the schools of Greek philosophy into the light of publicity, is his privilege. To hunt through musty and worm-eaten volumes for facts regarding the great physicians of antiquity is his delight; and to communicate the knowledge thus obtained to others, who have not the time or the facilities for such research, is his duty. In every period are events and incidents of interest, but to the Middle Ages a peculiar fascination attaches; for it was during this period that Europe, emerging from an intellectual darkness of ten centuries' duration, awoke to the Renaissance, and Medicine, as ever has been the case, kept pace with the general advance of knowledge.

The present book deals with the life of a master whose work was an essential factor in the evolution of the Anatomical Renaissance. In order to understand the New Birth of Anatomy it is necessary to know something of the scope and influence of the General Renaissance.

The General Renaissance

This, the Revival of Learning, includes an indefinite time in European history. The seeds of the new movement were planted in the Middle Ages, but they bore no fruit until the time had arrived for an apparently "spontaneous outburst of intelligence". Definitions of the Renaissance will vary with the point of view. Artists and sculptors will say it was a revolution which was created by the recovery of ancient statues; littérateurs and philosophers look upon it as a radical change due to the discovery of the writings of the classical authors; astronomers and physicists will cite the names of Copernicus, Galileo, and Torricelli; geographers will point to the discovery of a New Continent; historians will name the extinction of feudalism and the capture of Constantinople by the Turks; inventors will recall the changed conditions of warfare brought about by gunpowder, the multiplication of books by the invention of printing, and the advent of new methods of engraving; and anatomists will sound the praises of Leonardo da Vinci and of Andreas Vesalius. All will agree that the Renaissance meant Revolution—revolution in thought, in conduct, in creed, and in conditions of existence. To no one fact can the Renaissance be attributed; nor can its scope be limited to any one field of human endeavor. The Renaissance was, and is, and will continue to be, as long as the race progresses.

The new movement began in Italy and grew rapidly. When, toward the end of the sixteenth century, the lamp of learning began to get dim in Italy, it was relighted by the nations of northern Europe—the Germans, the Hollanders, and the English—and by them was transferred to us. The Revival consisted largely in the recovery of the buried writings of the ancient Greek and Roman authors, together with comments on what they had written, and the production of books which were modeled after their works. But it was broader than this. It included all branches of learning, although more progress was made in some lines than in others.

Italy, a country divided into numerous small States, and so-called Republics, offered great opportunities for individual development and became famous in those paths in which individualism has gained its greatest triumphs. Thus in literature, in law, in medicine, in painting and in sculpture, the Italians were preëminent. In architecture and in the drama they reached no such heights as were attained by the French, the Germans and the English. It was in the northwest part of Italy, in the province of Tuscany, that the Renaissance gained its greatest victories. Among the earliest of the leaders of the New Learning was the Florentine poet, Dante Alighieri (1265-1321). "To Dante", says Symonds, "in a truer sense than to any other poet, belongs the double glory of immortalising in verse the centuries behind him, while he inaugurated the new age". His *Vita Nuova* (New Life) and *Divina Commedia* (Divine Comedy) are essentially modern in thought, but ancient in the manner in which the thought is expressed.

Petrarch may be said to fairly open the new era. Like Dante, he was a Florentine. He was the apostle of Humanism, that system of philosophy which regarded man "as a rational being apart from theological determinations" and perceived that "classic literature alone displayed human nature in the plenitude of intellectual and moral freedom". To a revolt against the despotism of the Church, it added the attempt to unify all that had been taught and done by man. Petrarch was a poet, a lawyer, an orator, a priest, and a philosopher. He lived between the years 1304-1374. He was a great traveler, and visited the leading continental cities in order to converse with learned men. Petrarch delighted in the study of

Cicero, in collecting manuscripts, and in accumulating coins and inscriptions for historic purposes. He advocated public libraries and preached the duty of preserving ancient monuments. He opposed the physicians and astrologers of his day, and ridiculed the followers of Averröes.

Boccaccio, who has been called the Father of Italian Prose, and is most widely known as the author of the *Decameron*, did not spend all of his time in describing the escapades of the knights and ladies of old. Influenced potently by Petrarch, Boccaccio regretted the years he had wasted in law and trade, when he should have been reading the classics. Late in life he began the study of Greek that he might read the *Iliad* and the *Odyssey*. What he lacked in genuine scholarship he made up in industry. He continued the work begun by Petrarch of hunting for lost manuscripts of the ancient Greek and Roman authors. Many of these precious documents were stored in the conventual libraries, where, too often, they were either wantonly destroyed or were mutilated, the words of the author being erased from the parchment to make way for new prayers. Boccaccio tells of a visit which he made to the Benedictine Monastery of Monte Cassino near the city of Salernum. He wished to see the books and found them in a room without door or key. Many of them were mutilated. On making inquiry as to the cause, the monks answered that they had sold some of the sheets, having first erased the original words, replacing them with psalters. The margins of the old pages were made into charms and were sold to women.

It was owing to the unselfish labors of such men as Petrarch and Boccaccio that the works of Livy, Cicero, Quintilian, Terence, and others of the ancient authors, were preserved. In this enterprise they were encouraged by the rulers. Thus Cosimo de' Medici in Florence, Alfonso the Magnanimous in Naples, and Nicholas V. in Rome, to say nothing of the despots of the smaller cities, rivaled one another in their zeal in unearthing and multiplying the manuscripts of the ancient writers. They spared neither time nor money to increase their store of manuscript books. They surrounded themselves with learned men who lived in high esteem, and who were supported by salaries paid by the State or by private pensions.

The fifteenth century, which was one of the most remarkable epochs in history, was rich in accomplishment. Almost all of the great events which have influenced European commercial and intellectual development can be traced to that period. The invention of printing, [6] the discovery of America, the fall of the Roman Empire in the East, the birth of the Reformation, and the rise of art in Italy, all belong to this wonderful century. In this period, when almost every city in Italy was a new Athens, the Italian poets, historians, and artists vied with the eminent men of the ancient world in carrying the lamp of learning. The Italian cities—Florence, Bologna, Milan, Venice, Rome and Ferrara—fought with one another, not for the spoils of the battlefield but for the victories of science and of art; not so much for the profits of commerce as for the wealth of genius and of learning. The intellectual development which occurred in northern Italy under the rule of the house of Medici, and particularly under the auspices of Lorenzo the Magnificent, forms one of the most interesting periods in European history.

It is impossible in the present work to trace the steps by which the exquisite taste of the ancients in works of art was revived in modern times. Nevertheless, a few words may be devoted to this subject. While much must be credited to those Greek artists who had left their country and had settled in the Italian peninsula, it must be conceded that many of the works of art of the native Italians were not the less meritorious. The same circumstances which favored the revival of letters, operated to further the cause of art; and the same individuals, who were interested in the preservation of the manuscripts of the older authors, also busied themselves with the collection of ancient statues, paintings, gems and tapestry. The freedom of the Italian Republics permitted the minds of men to expand to full fruition; and the encouragement which was given by its rulers to artists, sculptors and artisans, made the city of Florence, in the fifteenth century, a not less renowned centre of culture than Athens had been in ancient times.

The revival of art dates from the time of Cimabue (1240-1300) and Giotto (1276-1336). The former is known as the Father of Modern Painters; the latter constructed the Campanile at Florence. To Giovanni Cimabue, scion of a noble Florentine family, is usually given the credit of being the restorer of art in Italy. He is thought to have been the first painter to throw expression into the human countenance. His work, if judged by present standards, would be called crude, rude and incomplete. Much of the fame of this painter is to be attributed to his being the first person whom Vasari chronicled in his *Lives of the Painters*.

For more than a century after the time of Cimabue and Giotto, painters displayed only a smattering of anatomical knowledge.

Early in the fifteenth century two Flemish artists, Hubert van Eyck (1365-1426) and his brother John (1385-1441), in their polyptych of the Adoration of the Lamb, boldly struck out along new lines and committed the unheard-of deed of painting nude figures. Italy, however, was the real birthplace of Art-Anatomy. While the Flemings and others of the North painted everything that they saw, including the nude, the Italians were the first men of the Renaissance who thought of painting the nude figure before draping it. Leo Battista Alberti (1404-1472), in his works on painting, insists that the bony skeleton must first be drawn and then clothed with its muscles and flesh. This was an important step in advance, since it shows that the Florentine artists were progressing towards realism and were breaking away from the symbolism of the early Christian painters and mosaic-workers. The new movement in art found a worthy champion in Antonio Pollaiuolo (1432-1498). In his knowledge of the anatomy of the human figure he surpassed all of the artists of his day; and as a result of his labors he may justly be named the founder of the scientific study of the nude. His knowledge of anatomy was so accurate, and so extensive, that it could have been gained only in the dissecting room.

Under the patronage of Lorenzo de' Medici and the guiding mind of Pollaiuolo, there occurred a revival of pseudo-paganism in Art. The old Church subjects were largely neglected; mythological subjects again became the fashion; draperies were either modified or were laid aside; and the scientific study of anatomy, both as regards the nude figure and the dissection of the individual parts, became the necessary training of the student. Of all the masters of this period, the palm for excellence in drawing the naked figure must be awarded to Luca Signorelli (1442-1524), from whose work Michael Angelo is known to have profited.

The alliance between skilled anatomists and master artists was of reciprocal benefit. The anatomical studies which were made conjointly by Leonardo da Vinci and the celebrated teacher of anatomy, Marc Antonio della Torre, were lost to the world by the untimely death of the latter, before he had finished a magnificent treatise on human anatomy. Leonardo's anatomical sketches, if they had been published during his lifetime, would have revolutionized anatomy both as regards discoveries in the body and the teaching of the structure of man. These masterpieces of anatomical illustration long remained hidden from the world; they were published only in the year 1902. Even now their cost is so great that only a few wealthy libraries can possess them. Leonardo's long unpublished drawings show him to have been a most accurate anatomist. At the same time, he constantly kept in view the aim of fine art, which, in so far as practical anatomy is concerned, needs a knowledge of only the bones and the muscles.

Nor was Leonardo the only artist who made dissections. Raffaello Santi, Michael Angelo, Bartholomaus Torre, Luigi Cardi or Civoli, Jan Stephan van Calcar, Giuseppe Ribera, Arnold Myntens, and Pietro da Cortona studied practical anatomy. Rubens's long-lost sketch-book[1], which was published one hundred and thirty-three years after his death, shows with what care he had studied human anatomy. Albrecht Dürer's *Treatise on the Proportions of the Human Body* is also worthy of mention.

In the number and fame of her Universities, Italy showed supremacy. At the end of the fifteenth century she could boast of sixteen seats of learning, a number equal to that of the combined institutions of Britain, France, Germany, Hungary, Bohemia and Bavaria.

This digression has led us away from the Humanists. Their list is a long one. Among them were Poggio Bracciolini, who discovered the manuscript of the *Institutions* of Quintilian and the writings of Vitruvius; Poliziano, the first poet of the fifteenth century, and the translator of the works of Hippocrates and Galen; Pontanus, whose *De Stellis* and *Urania* were much admired by Italian scholars; Sannazzaro, whose epic on the birth of Christ cost him twenty years of labor; Vida, whose *Christiad* and other poems were much admired; and Fracastoro, whose *Syphilis* was hailed as a divine poem.

From the viewpoint of the medical historian an important event occurred in the year 1443, when Thomas of Sarzana, later known as Pope Nicholas V., discovered a manuscript copy of the *De Medicina* of Aulus Cornelius Celsus. This classic, which had been lost for many centuries, was one of the first medical books to pass through the press. It gave physicians an insight into Hippocratic medicine without the disadvantage of an imperfect translation. Physicians took an active part in the Renaissance. Thus Nicholas Leonicenus, of Ferrara, translated the *Aphorisms* of Hippocrates and the *Natural History* of Pliny; and Winter of Andernach did similar labor for the writings of Galen, Alexander, and Paulus

Aegineta. Their efforts seem insignificant in comparison with those of Anutius Foesius, a humble practitioner of Metz, who spent forty years of his life in preparing a complete Greek edition of the works of Hippocrates. The New Learning was brought to England by two physicians, Thomas Linacre and John Kaye (Caius).

Some of the Humanists were printers. The history of printing in Italy naturally forms a part of the history of the Renaissance. In 1462, Maintz was pillaged by Adolph of Nassau and its printers were scattered over Europe. Two of them wandered into Italy, living in a village in the Sabine mountains, where, in October, 1465, the first book was printed from an Italian press. It was a Latin edition of Lactantius. Six years later a press was established in Florence. In 1478, Mondino's *Anathomia* was printed in Pavia. It has been estimated that before the first year of the sixteenth century, five thousand books had been printed in Italy. In those days the editions were small, 265 copies being considered one edition. An immense amount of labor was required to get out a new edition. First, the manuscripts of the ancient author had to be collected, compared and corrected, this work being done by learned men who resided in the home of the publisher. The corrections were made without the aid of dictionaries, grammars, or book-helps of any kind. The proof was read aloud to the assembled scholars and the final corrections were added. In time, Venice came to be the most noted of the Italian cities in the publishing business, owing chiefly to the family of Aldo. This family of printers became famous for finely printed Greek and Latin books, which are still called Aldine editions. Nine years after the printing of the first book in Italy, the art was practiced in England by Caxton.

Humanism in Italy began to decline toward the close of the fifteenth century. Long before this time it had degenerated into Paganism. The scholars influenced all life, customs and thought. Although the nation remained Catholic, it was such only in name. Everyone bowed before the shrine of classical literature. Even in the christening of children the Christian name was sacrificed to paganism. The saints were forgotten, and the names most frequently chosen were those from heathen mythology. The polite authors described scenes, events and actions in their writings in terms which long since have been banished from good society. A spade was called by its true name. Bembo, the secretary of Leo X., could write a hymn to Saint Stephen or a monologue for Priapus with equal ease and elegance. The amours of the high and the low were flaunted in print. The nation degenerated into an intellectual and sensual state which involved even the Popes. Scholars and rich men alike vied with one another in returning to those pursuits, habits, and methods of thought which had ruled ancient Rome in her most corrupt days.

Such a condition could not exist forever. The turning-point came in 1527, when Charles the Fifth, engaging in war with Pope Clement VII., captured and sacked the city of Rome. After that event everything was changed. Not only had the scholars lost their influence, but many of them had lost their lives. Valeriano, who returned to Rome after the siege, pathetically exclaims: "Good God! when first I began to enquire for the philosophers, orators, poets and professors of Greek and Latin literature, whose names were written on my tablets, how great, how horrible a tragedy was offered to me! Of all those lettered men whom I had hoped to see, how many had perished miserably, carried off by the most cruel of all fates, overwhelmed by undeserved calamities; some dead of plague, some brought to a slow end by penury in exile, others slaughtered by a foeman's sword, others worn out by daily tortures; some, again, and these of all the most unhappy, driven by anguish to self-murder". Such was the end of the men who made the Italian Renaissance. The Spaniards, the Inquisition, and the changed policy of the Church prevented a second revival of Humanism.

While the sack of Rome marks the end of the Humanists, the Revival in Medicine continued to grow in vigor and extent. Many of the greatest discoveries in anatomy were made, and most of the important books on this subject were written, in the middle and latter part of the sixteenth century. Italian history is rich in contradictions. While peace, ease and comfort are generally considered to be necessary to the development of science and culture, Italy offers the strange spectacle of the steady increase in medical knowledge in spite of wars and alarms. The Inquisition, which had been introduced from Spain in 1224, was given a new and horrible impetus when, in 1540, Paul III. appointed six cardinals to add to its tortures. One of them, Caraffa, became Pope Paul IV. in 1555, and four years later originated the *Index Expurgatorius*. Torn by civil and foreign wars, and terrorized by the Inquisition, which was not abolished until late in the eighteenth century, Italy gradually lost her commercial and intellectual supremacy. That she should have accomplished so much under such unfavorable circumstances, is now a matter of wonderment.

The origin of the Renaissance in Italy was due to many causes. The early Roman civilization was not entirely blotted out by the invasion of the barbarians of the North. And in the matter of language the Italians possessed an advantage, since the transition from Latin to Italian was easier than from Latin to Spanish, French, English or German. The fertility of the country; the mildness of the climate; the division into semi-independent states; the infusion of new northern blood into the veins of the Italians; the removal of the papal court to Avignon in 1309; and the gradual rise of a powerful middle class, whose members included the devotees of the professions of law and medicine, were factors which determined that Italy, rather than France or Spain, should be the field for the Revival of Letters.

To Italy, then, belongs the glory of having been the first to free herself from the trammels of ancient scholasticism and the fetters of mediaeval theology. She abandoned the wordy dialectics and metaphysical gymnastics of the philosophers of old. In place of mortification, penance and solitary confinement in cloistered monasteries and convents, she began to have a proper conception of the dignity of man and his relation to nature.

Italy, in the time of her freedom, received the torch of learning from Greece; Italy revived its brilliancy, and, when her time of adversity and ruin arrived, she passed it on to the nations of Northern Europe. They in turn have transferred it to America, to Australia, to India, and to the uttermost parts of the earth.

The Anatomical Renaissance

Italy in the sixteenth century was the fount from which issued a ceaseless stream of anatomical discoveries. The ancient and illustrious Universities of Bologna, Pavia, Padua, Pisa and Rome, eclipsed the schools of Paris and Montpellier, of Toulouse and Salamanca; and the Italian peninsula, which, in early mediaeval times, had gloried in the skill of the physicians of Salernum, a second time became the medical centre of Europe. Vesalius and his pupil, Fallopius, taught at Padua; the ancient fame of Bologna was supported by Arantius and Varolius; Vidius, returned from establishing the anatomical school at Paris, taught at Pisa; Eustachius was at Rome, Ingrassias lectured at Naples, and the fame of the New Anatomy spread throughout the world. The Italian cities were filled with students from foreign lands. Padua had more than one thousand new students every year, salaries were paid to her one hundred professors, and medicine was looked upon as a noble profession.

While the Italians were the leaders in progress, the Germans were still lecturing on Galen and Avicenna, the English had done almost nothing, and the Collége de France was not established until 1530.

Legalized by imperial authority and sanctioned by the Church, dissection was no longer regarded as a crime. A bull by Pope Boniface VIII., issued in the year 1300, forbidding the evisceration of the dead and the boiling of their bodies to secure the bones for consecrated ground, as was done by the Crusaders, was wrongly interpreted as forbidding anatomical dissection. Two centuries later the Popes, standing in the vanguard of science, permitted dissections to be made in all the Italian medical schools, and paved the way for the Anatomical Renaissance.

Great things were done in the sixteenth century. Under the scalpel and pen of Vesalius, anatomy was revolutionized. Surgery was guided into new paths by Ambroise Paré; and obstetrics, thanks to the labors of Eucharius Rhodion and Jacques Guillemeau, began to assume its legitimate place among the medical sciences. Servetus, visionary and argumentative, correctly described the pulmonary circulation in a theological work which was burned with its author. Eustachius, Columbus and Fallopius widened the path which had been blazed by Vesalius. Arantius, Caesalpinus and Fabricius added materially to anatomical science. The labors of all these great masters prepared the way for the greatest event occurring in the seventeenth century, namely, William Harvey's discovery of the circulatory movement of the blood.

INITIAL LETTER BY VESALIUS
(From the "Fabrica", 1543)

CHAPTER FIRST
Anatomy in Ancient Times

Egypt and Greece were the sources of the medical learning of the ancient world. Although the Egyptians and early Greeks possessed a certain amount of anatomical knowledge, which was gained in the one instance by the practice of embalming and in the other by an examination of the bones, no real progress could be made because of the laws, customs and prejudices of those ancient peoples. Thus we find the Egyptians stoning the operator who opened the abdomen in order that the body might be embalmed; and the Greeks inflicted the death penalty on those of their generals who, after a battle, neglected to bury or burn the remains of the slain.

HIPPOCRATES

In the time of Hippocrates, whose life extended approximately over the period between 460-377 B.C., Greek medicine emerged from the domination of the Asclepiadae, or priests of Aesculapius, who had followed it as an hereditary and secret art. Prior to this time in the numerous Asclepia, or Temples of Aesculapius, votive offerings had been accepted, some of which were of anatomical interest. Thus the Temple at Athens received a silver heart and gold eyes. Pausanias states that Hippocrates gave to the Temple of Apollo, at Delphos, a skeleton which was made of brass. Possibly, as Moehsen[2] believes, this was a metallic figure representing a man who was much emaciated by the ravages of disease. In the Hippocratic writings, some of which are undoubtedly spurious, are few references to the opening of a dead body; and these examinations concern the investigation of the thorax and abdomen in order to determine the cause of death. While the Greek physicians knew little of the human muscles, of the nervous system and of the organs of sense, they were well acquainted with the anatomy of the bones. Their dissections were held upon the lower animals.

It is impossible to determine whether or not the Greek physicians of the Hippocratic period dissected the human body. "It has long been a matter of debate", says John Bell[3], "whether the ancients were, or were not, acquainted with anatomy, and the subject, with its various bearings, has been much and keenly agitated by the learned. If anatomy had been much known to the ancients, their knowledge would not have remained a subject of speculation. We should have had evidence of it in their works; but, on the contrary, we find Hippocrates spending his time in idle prognostics, and dissecting apes, to discover the seat of the bile."

Galen[4] states that the ancient physicians did not write works on anatomy; that such treatises were at that time unnecessary, because the Asclepiadae—to which family Hippocrates belonged—secretly instructed their young men in this subject; and that opportunities were given for such study in the temples of Aesculapius.

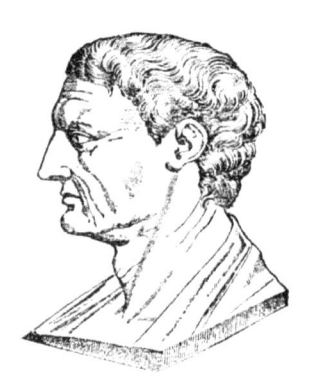

ARISTOTLE

The first systematic dissections seem to have been made by the Pythagorean philosopher Alcmaeon, who lived in the sixth century B. C., but it is uncertain whether he dissected brutes or men. The cochlea of the ear and the amnios of the foetus were named by Empedocles of Agrigentum, in the fifth century B. C. The nerves were first distinguished from the tendons by Aristotle, (384-322 B. C.), the most celebrated zoötomist of antiquity, who has been called the Father of Comparative Anatomy. For twenty centuries his views of natural phenomena were held in high esteem.

For a long period the early inhabitants of Rome were practically without physicians. During severe epidemics they had recourse to oracles, to the health deities of the Greeks, and to their native gods. As early as the fifth century B. C., during a pestilence, a temple was erected to Apollo as Healer. The worship of Aesculapius was introduced into Rome in the year 291 B. C. Livy relates that the god of medicine in the guise of a serpent was transported from Epidaurus, in Greece, to the Isle of the Tiber where a temple was built in his honor.

The Romans, like the Greeks, were accustomed to leave votive offerings, or donaria, in their temples. Such gifts included surgical instruments, pharmaceutical appliances, painted tablets representing miraculous cures, and great numbers of images of various parts of the human frame shaped in metal, stone or terra-cotta. Among the remains of Roman anatomical art is the marble figure which was unearthed in the villa of Antonius Musa, the favorite physician of the Emperor Augustus. It is a human torso; the front of the chest and abdomen has been removed so as to expose the viscera. The heart is placed vertically in the middle of the thorax, thus corresponding to the position of this organ as described by Galen who made his dissections on apes. It is a human thorax with simian contents. The figure is supposed to have been constructed for the purposes of a teacher of anatomy.

ALEXANDER THE GREAT

It was in the famous Alexandrian University that human anatomy was first studied systematically and legally.

Alexander the Great, after the fall of Tyre (332 B. C.) and the siege of Gaza, ordered his fleet to sail up the Nile as far as Memphis while he proceeded overland with the army. It was probably on this march, while viewing the pyramids and other marvelous works of the ancient Egyptians, that he conceived the grand idea of founding a city upon the banks of the Nile, which should be a model of architectural beauty, a centre of intellectual life and a lasting monument of his own greatness and magnificence. The foundation of Alexandria was laid by the warrior whose name it bears; but the credit of instituting the Library belongs to one of his lieutenants, Ptolemy Soter.

PTOLEMY SOTER

The new city which for centuries was the intellectual and commercial storehouse of Europe, Africa and India, was of oblong form. Lake Mareotis washed its walls on the south, while the Mediterranean bathed its ramparts on the north. Provided with broad streets, it was adorned with magnificent houses, temples and public buildings. At the centre of the city was the Mausoleum in which was deposited the body of Alexander, embalmed after the manner of the Egyptians. Alexandria was divided into three parts: the *Regio Judaeorum* or Jews' quarter, in the northwest; the *Rhacotis*, or Egyptian section, on the west, containing the Serapeum with a large part of the Library; and on the north, the *Bruchaeum*, or Greek portion, containing the greater part of the Library, the Museum, the Temple of the Caesars and the Court of Justice. The population was cosmopolitan in character; the statues of the Greek gods stood by the side of those of Osiris and of Isis; the Jews forgot their language and spoke Greek; and under the Ptolemies, who were of Greek descent, Alexandria became a centre of intellectual life and culture.

To the medical historian the most interesting feature of Alexandria was the Museum or University. Here were assembled the intellectual giants of the earth: Archimedes and Hero, the philosophers; Apelles, the painter; Hipparchus and Ptolemy, the astronomers; Euclid, the geometer; Eratosthenes and Strabo, the geographers; Manetho, the historian; Aristophanes, the rhetorician; Theocritus and Callimichus, the poets; and Erasistratus and Herophilus, the anatomists, all of whom labored in quiet upon the peaceful banks of the Nile. The early Christian church drew from "the divine school at Alexandria" such eminent teachers as Origen and Athanasius. Here were a chemical laboratory, a botanical and zoölogical garden, an astronomical observatory, a great library, and a room for the dissection of the dead.

In the Alexandrian school of medicine Erasistratus and Herophilus taught the science of organization from actual dissections. The generosity of the Ptolemies not only furnished them with an abundance of dead material, but condemned malefactors were used for human vivisection. Celsus[5] states that the Alexandrian anatomists obtained criminals, "for dissection alive, and contemplated, even while they breathed, those parts which nature had before concealed."

Herophilus made many anatomical discoveries. He traced the delicate arachnoid membrane into the ventricles of the brain, which he held to be the seat of the soul; and first described that junction of the six cerebral sinuses opposite the occipital protuberance, which to this day is called the *torcular Herophili*. He saw the lacteals, but knew not their use, and regarded the nerves as organs of sensation arising from the brain; he described the different tunics of the eye, giving them names which are still retained; and first named the duodenum and discovered the epididymis. He attributed the pulsation of arteries to the action of the heart; the paralysis of muscles to an affection of the nerves; and first named the furrow in the fourth cerebral ventricle, calling it *calamus scriptorius*.

Erasistratus gave names to the auricles of the heart; declared that the veins were blood-vessels; and the arteries, from being found empty after death, were air-vessels. He believed that the purpose of respiration was to fill the arteries with air; the air distended the arteries, made them beat, and in this manner the pulse was produced. When once the air gained entrance to the left ventricle, it became the vital spirits. The function of the veins was to carry blood to the extremities. He is said to have had a vague idea of the division of nerves into nerves of sensation and of motion; to the former he assigned an origin in the membranes of the brain, while the latter proceeded from the cerebral substance itself. He recognized the use of the trachea as the tube which conveys air to the lungs. A catheter, the first invented, which was figured in ancient surgical works, bore the name of the catheter of Erasistratus. He gravely tells us, as the result of his anatomical studies, that the soul is located in the membranes of the brain.

The practice of human dissection did not long exist in the city of its origin, and after the second century was unknown. Then science underwent a retrogression; observations and experiments were replaced by useless discussions and subtle theories. The decline of the Alexandrian University was due to a series of disasters which began with the Roman domination and reached their climax with the capture of the city by the Arabs.

GALEN

Claudius Galenus, the celebrated Roman physician whose writings were for centuries accepted as authority and whose reputation was second only to that of Hippocrates, was obliged to base his anatomical treatises largely upon the dissection of the lower animals. He advised his pupils to visit Alexandria, where he had studied, in order that they might examine the human skeleton. He complained that the physicians of his time—in the reign of Marcus Aurelius—had entirely neglected anatomical knowledge and had degenerated into mere sophists. He appreciated the importance of anatomy, particularly to a surgeon who is called upon to treat wounds and injuries. Hence he has endeavored in the four books, *De Anatomicis Administrationibus*, to cover this part of anatomy as exhaustively as possible.

Galen's voluminous writings form a precious monument of ancient medicine. The works of the Alexandrian anatomists having been destroyed, we know of their labors chiefly from what Galen has said of them. His treatises show a remarkable familiarity with practical anatomy, although his dissections were made upon the lower animals. Galen's knowledge of osteology was extensive. He described the bones of the skull, the cranial sutures, and the essential features of the malar, maxillary, ethmoid and sphenoid bones. He divided the vertebrae into cervical, dorsal and lumbar classes. He knew that both arteries and veins were blood-carrying vessels; he described the valves of the heart, and recognized this organ as the source of pulsation. He erroneously taught that the interventricular septum presents foramina through which the two kinds of blood become mixed.

In myology Galen made numerous advances. "Previous to his investigations", says Fisher[6] "much confusion existed as to what constituted a single muscle; he adopted the general rule of considering each bundle of fibers that terminates in an independent tendon to be one muscle. He was the first to describe and give names to the platysma myoides, the sterno- and thyro-hyoides, and the popliteal. He described the six muscles of the eye, two muscles of the eyelids, and four pairs of muscles of the lower jaw—the temporal to raise, the masseter to draw to one side, and two depressors, corresponding to the digastric and internal pterygoid muscles. He described also the brachialis anticus, the biceps flexor cubiti, the sphincter and levator ani, and the straight and oblique muscles of the abdomen. In short, he described the greater portion of the muscles of the body, his treatise differing chiefly from a modern one in the minute account of these organs and in the omission of some of the smaller muscles." Galen studied the brain and named the corpus callosum, the septum lucidum, the corpora quadrigemina and the fornix; but erroneously stated that the nerves of sensation arise from the brain, and those of motion from the spinal cord. He denied the decussation of the optic nerves. He described the pneumogastric and sympathetic nerves; seven pairs of cerebral and thirty pairs of spinal nerves; and claimed the discovery of the ganglia of the nervous system. He located the seat of the soul in the brain, which also is the source of the rational mind; the heart to him was the source of courage and of anger, and the liver was the seat of desire. Many of Galen's anatomical statements show that he derived his knowledge from comparative dissections.

The Galenic era was followed by that long period of ignorance, of slumber and of inaction which is justly known as the Dark Ages. While a few Greek and Arab writers, who came after Galen, contributed to the literature of medicine and surgery, they did nothing for anatomy. After the end of the fifth century even the works of Galen were forgotten. At this period, when medicine was chiefly in the hands of the Jews, the Arabs and the bigoted clergy, nothing was done for science or for art. The whole influence of Christianity was exerted against the schools of philosophy. Illustrious apostles of the Church pronounced anathemas against the reading of the ancient classics;[7] and eminent ecclesiastics regarded disease as a divine penalty or as an invaluable aid to saintly advancement. Art and anatomy were practically forgotten. Their Renaissance occurred almost simultaneously.

During the period from the seventh to the fourteenth centuries the school of Salernum was for medicine what Bologna became for law and Paris for philosophy. Here, for eight hundred years, medicine was taught to thousands of students and the impress of the profession was so potent that the city called itself *Civitas Hippocratica*, and thus its seals were stamped. Here medical diplomas were first issued to waiting students who took a sacred oath to serve the poor without pay. Here with a book in his hand, a ring on his finger and a laurel wreath on his head, the candidate was kissed by each professor and was told to start upon his way. Here women were professors and vied with men in spreading the doctrines of our art.

For a period of several hundred years anatomy was taught at Salernum from dissections made upon pigs. Copho, one of the Salernian professors of the early part of the twelfth century, wrote a treatise, *Anatomia Porci*, which gives minute directions regarding the manner in which the animal is to be dissected. Another anatomical work of later date, written by a member of the Salernian faculty, is entitled *Demonstratio Anatomica*; it also deals only with comparative anatomy. In the thirteenth century (A. D. 1231) Frederick II., Emperor of Germany and King of the Two Sicilies, and the author of a treatise which contained a complete anatomy of the falcon, decreed that a human body should be anatomized at Salernum at least once in five years. Physicians and surgeons of the kingdom were required to be present at the dissection. So far as is known, no record has been kept of these demonstrations. Creditable as was this anatomic decree, the great Hohenstaufen in other respects was not free from the errors of his age. A firm believer in *Medicina Astrologica*, he did not decide upon any undertaking until the stars had been consulted.

It was not alone at Salernum that dissection was legalized in the thirteenth century. A document of the year 1308, of the Maggiore Consiglio of Venice, shows that a medical college located in that city was authorized to dissect a body once a year. This, and other isolated examples, indicate that the time was approaching when anatomy should be taught from human dissections. The credit of reinaugurating the teaching of this useful department of science belongs to Mondino dei Luzzi of Bologna.

CHAPTER SECOND
Mondino, the Restorer of Anatomy

In the year 1315, in the old Italian city of Bologna, an event occurred which marks an important epoch in the history of medicine. A wondering crowd of medical students witnessed the dissection of a human cadaver—one of the few procedures of the kind that had occurred since the fall of the Alexandrian University. Acting under royal authority Mondino, a man far in advance of the age, placed the body of a female upon a table where for many centuries before only the cadavera of apes, of swine and of dogs had been studied.

Mondino, known also as Mundinus, Mundini, Raimondino, or Mondino dei Luzzi, was descended from a prominent Italian family. Little is known of his life. The year of his birth is disputed; probably 1276 was near the time. He was graduated in medicine in 1290 and in 1306 he became a professor in the University of Bologna, holding his chair with credit until his death in 1326. Like that of the illustrious Homer, Mondino's nativity has been claimed by several rival cities. Guy de Chauliac, writing in 1363, states that Mondino was a Bolognese: *Mundinus Bononiensis* is Chauliac's expression.

Mondino's method of teaching anatomy is known from Chauliac's testimony:—"Mundinus of Bologna, wrote on anatomy, and my master, Bertruccius, demonstrated it many times in this manner:—The body having been placed on a table, he would make from it four readings; in the first the digestive organs were treated, because more prone to rapid decomposition; in the second, the organs of respiration; in the third, the organs of circulation; in the fourth the extremities were treated." The innovation so auspiciously begun was not continued, and after the death of Mondino human dissections were made only at long intervals. The few instances in which, in the fourteenth and fifteenth centuries, the ecclesiastical and civil authorities granted the right to make dissections only prove the contention, that the practical study of human anatomy did not gain recognition until the sixteenth century.

When Mondino began his dissections the epoch of Saracen learning had ended, but the influence of Arab medicine exerted by the writings of Albucasis, Avicenna and Rhazes had not declined. The Arabian physicians had accomplished little for anatomy. In this line the influence of Galen was still potent, and was rarely questioned until the publication of the *Fabrica* of Vesalius in 1543. During a long period the little treatise of Mondino held full sway in the mediaeval schools. Medicine was taught in the University of Bologna, which as early as the twelfth century was celebrated for its departments of literature and of law. These studies were free of the difficulties which beset medicine. The prejudice against dissection was so great that for nearly a century after his death few men dared to repeat the acts of Mondino.

In 1316 Mondino issued his book which remained in manuscript form for more than one hundred and fifty years, the first printed edition bearing the date 1478. Small and imperfect as it was, it marks an era in the history of science. By command of the authorities this book was read in all the Italian Universities. The work of Mondino contained no new facts; it was compiled largely from the writings of Galen and of Avicenna. The descriptions, to use the words of Turner, "are corrupted by the barbarous leaven of the Arabian schools, and his Latin is defaced by the exotic nomenclature of Ibn-Sina and Al-Rasi". Mondino divided the body into three cavities, of which the upper contains the animal members, the lower the natural members, and the middle the spiritual members. Many of his names are borrowed from the Arab writers. Thus, he calls the peritoneum *siphac*, the omentum *zyrbi*, and the mesentery *eucharus*. His description of the heart is much nearer accuracy than would be expected. He resorted to vivisection, and tells us that when the recurrent nerves of the larynx are cut the animal's voice is lost. In his book we find the rudiments of phrenology. He states that the brain is divided into compartments, each of which holds one of the faculties of the intellect.

MONDINO'S DIAGRAM OF THE HEART, 1513

Mondino did not himself make the dissections which are credited to him. According to an ancient custom which lasted until the time of Vesalius, the actual cutting was done by a barber who wielded a knife as large as a cleaver. The professor of anatomy sat upon an elevated seat and discoursed concerning the parts, while a demonstrator, who also did not soil his fingers, pointed to the different structures with a staff. Originally Mondino's book contained no figures; when the art of wood engraving was introduced in the latter part of the fifteenth century, a few rude woodcuts appeared which represent Mondino and his method of teaching. In the *Fasciculus Medicinae* of Joannes de Ketham, published at Venice in 1493, Mondino's book is printed with an illustration showing a demonstration in anatomy.

According to Mondino the heart is placed in the centre of the body. The valves he considers "wonderful works of nature". He describes a right, left and middle ventricle. The right ventricle has thinner walls than the left, because it contains blood; the left one contains the vital spirit, which passes through the arteries to the body; and the middle ventricle consists of many small cavities "broader on the right side than on the left, to the end that the blood, which comes to the left ventricle from the right, be refined, because its refinement is the preparation for the generation of vital spirit, which should be continually formed". Mondino describes five bones of the head, separated by three sutures—coronal, sagittal and occipital. The brain has two membranes: dura and pia. There are three cerebral ventricles—anterior, posterior and middle—and in these he locates the various intellectual qualities. He describes the cerebral nerves: olfactory, optic, motor oculi, facial, vagus, trigeminal, auditory and hypoglossal. He calls the innominate bone *os femoris*: the femur, *canna coxae*; the humerus, *os adjutori*; while the bones of both leg and forearm are named *focilia* major and minus.

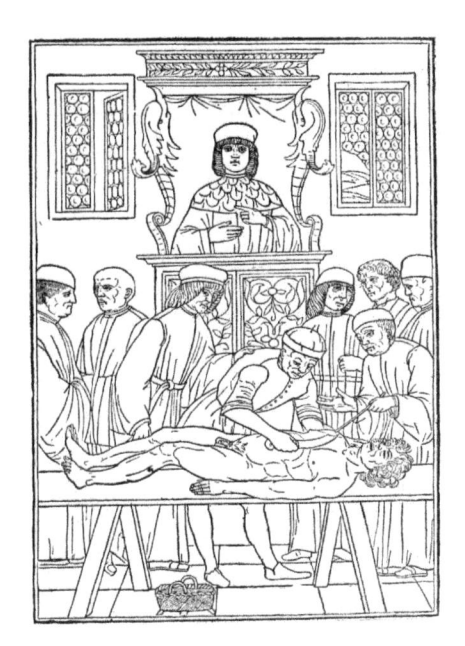

ANATOMICAL DEMONSTRATION IN 1493
(Joannes de Ketham)

TITLE-PAGE OF MONDINO'S ANATOMY BY
MELERSTAT
(Printed before 1500)

Like many anatomists who succeeded him, Mondino mingled surgical ideas with his anatomical statements. A break in the *siphac* causes hernia and a swelling in the *mirach*. He treated ascites by puncture and evacuation, making a valve-like opening. Wounds of the large intestines must be sutured; if the wound be in the small intestines he advises that "you should have large ants, and, making them bite the conjoined lips of the wound, decapitate them instantly, and continue until the lips remain in apposition and then reduce the gut as before". He gives an explanation of the length and convolution of the intestines; "for if it were not convoluted the animals would have to be continuously ingesting food and continuously defecating, which would impede engagement in the higher occupations". Digestion is aided by black bile from the spleen and by red bile from the liver. The kidneys he regards as glands in which urine is extracted from the blood. The renal veins expand and form a fine membrane like a sieve through which the urine is filtered but blood cannot pass. He mentions renal calculi: if small they pass through the ureter; if large they are incurable except by incision, and this is to be avoided. The uterus and breasts are connected by veins, hence the sympathy between these organs. Inguinal hernia is to be operated upon; the spermatic cord and testicle may or may not be dissected out, or the hernia may be treated by the application of a caustic. An incision in the neck of the bladder will heal, because this part is muscular; but a cut in the body of the organ will not heal. He describes the operation for stone:—The patient being in proper position, the stone is conducted to the neck of the bladder by the finger in the rectum; an incision is made and the stone is pulled out with an instrument called *trajectorium*.

Mondino's book passed through not less than twenty-three editions between the years 1478-1580. The only manuscript extant is in the National Library at Paris.

The first printed edition of the *Anathomia Mundini*, Pavia, 1478, is a folio of twenty-two leaves. The Strassburg edition, 1513, is a small octavo volume of forty leaves. It contains a diagram of the heart and an astrological figure, a cadaver with the thorax and abdomen opened, surrounded by the signs of the zodiac. Such was the volume which for more than two hundred years was supposed to contain all that was to be said of human anatomy!

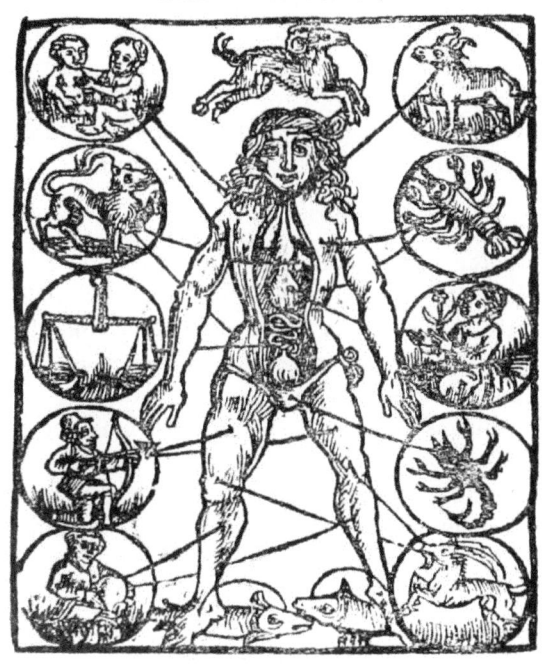

COLOPHON OF THE ANATOMY OF MONDINO, 1513

So numerous are the abbreviations in Mondino's book, so barbarous is his style, that the making of a translation is a difficult task. His reasons for writing are these:—"A work upon any science or art—as saith Galen—is issued for three reasons; *First*, that one may help his friends. *Second*, that he may exercise his best mental powers. *Third*, that he may be saved from the oblivion incident to old age".

CHAPTER THIRD
Mondino's Successors

For two hundred years anatomists used Mondino's book as a text for their lectures and for the same period anatomical writers did little more than comment upon this treatise. The new art of wood engraving was turned to anatomical use and crude illustrations of the various parts of the body were put into circulation. Some of these pictures were in the form of *Fliegende Blätter*, or flying leaves. A set of anatomical plates of this type was issued by a certain Ricardus Hela, a physician of Paris, as early as the year 1493. They were printed at Nuremberg. Their character may be judged by the accompanying illustration of the osseous system.

Gabriel de Zerbi

One of Mondino's commentators was Gabriel de Zerbi (1468-1505), of Verona, who taught medicine, logic and philosophy in the Universities of Padua, Bologna and Rome. His book, *Anatomia Corporis Humani*, appeared at Venice in 1502. Zerbi imitated Mondino in style, abbreviations and language. The work, however, contains some original observations regarding the Fallopian tubes, the puncta lachrymalia and the lachrymal gland. From the fact that Zerbi describes two lachrymal glands in each orbit, it is known that many of his dissections were made upon brutes.

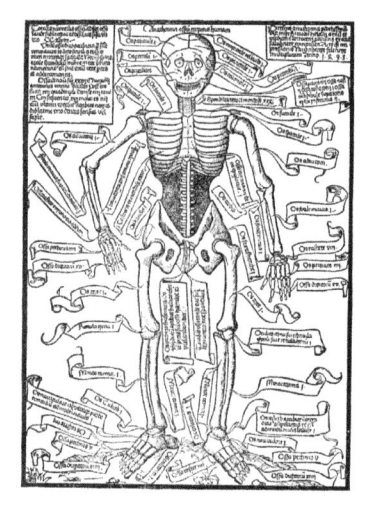

ANATOMICAL PLATE BY RICARDUS HELA, 1493

Zerbi's reputation, which extended to all parts of Europe, was the cause of his death. The Venetians received from Constantinople the request for a skillful physician who should treat one of the principal Seigniors of Turkey. The Republic turned its eyes to Zerbi who went to Constantinople, apparently cured the Seignior, and, loaded with presents, started on the return voyage for Venice, Unfortunately the patient suddenly died after a debauch. The infuriated Turks overtook the ship on which Zerbi and his son were passengers and carried them back to Constantinople, where both the anatomist and his son were quartered alive.

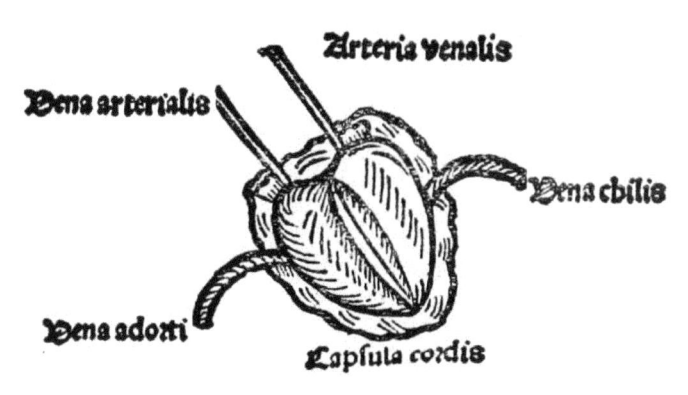

PEYLIGK'S DIAGRAM OF THE HEART, 1499

John Peyligk

Among the German anatomists of this period was John Peyligk, a Leipsic jurist, whose *Philosophiae Naturalis Compendium*, printed at Leipsic in 1499, contains crude anatomical illustrations.

Magnus Hundt

Far more important was the *Antropologium* of Magnus Hundt (1449-1519), of Magdeburg, which appeared at Leipsic in 1501. It contains four large and several small woodcuts which are among the earliest of anatomical illustrations. One of these shows the trachea on the right side of the neck, passing downward to the lungs; on the left side the oesophagus is represented. In the thorax are seen the lungs and the heart, the latter resembling the figure of this organ as presented on old playing cards. The pericardium has been opened, and the stomach and intestines are crudely figured. The diaphragm is absent.

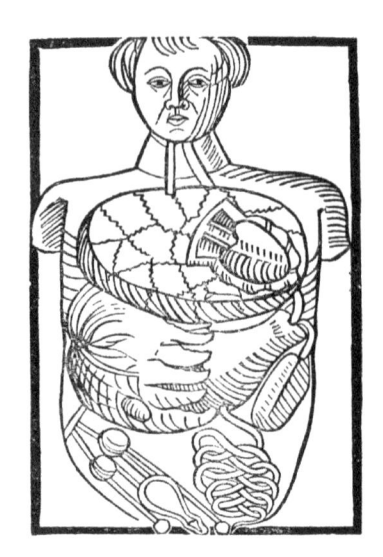

ANATOMICAL FIGURE FROM MAGNUS HUNDT, 1501

Laurentius Phryesen

Early in the sixteenth century a Holland physician, Laurentius Phryesen (*Phries, Friesen*), residing in the German city of Colmar and later at Metz, wrote a popular book on medicine, *Spiegel der Artzny*, which was published at Strassburg in 1518. It contains two anatomical illustrations cut in wood, dated 1517, and supposedly made after the drawings of Waechtlin, a pupil of the Elder Holbein. These pictures tell their own story; they show a marked improvement over the figures which Hundt published in 1501. The other anatomical plate in Phryesen's book is devoted to the skeleton.

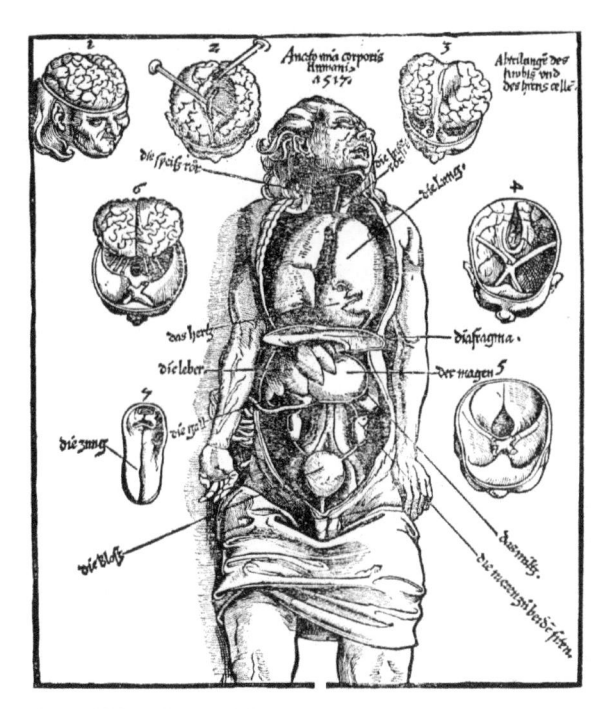

ANATOMICAL FIGURE FROM LAURENTIUS
PHRYESEN, 1518

Alexander Achillinus

The Italian physician Alexander Achillinus (1463-1525), professor of philosophy and medicine in Bologna, is deserving of mention for his anatomical knowledge. Zealously devoted to the Arab medical authors, Achillinus made numerous discoveries which are set forth in his general anatomy, *De Humani Corporis Anatomica*, Venice, 1516; and in a commentary upon Mondino's book, *In Mundini Anatomiam Annotationes*, Venice, 1522. He discovered the duct of the sublingual gland, usually credited to Wharton; two of the auditory ossicles, the malleus and incus; the labyrinth; the vermiform appendix; the caecum and ileo-caecal valve; and the patheticus nerve. Portal credits him with a better knowledge of the bones and of the brain than was possessed by his predecessors.

ALEXANDER ACHILLINUS

Berengario da Carpi

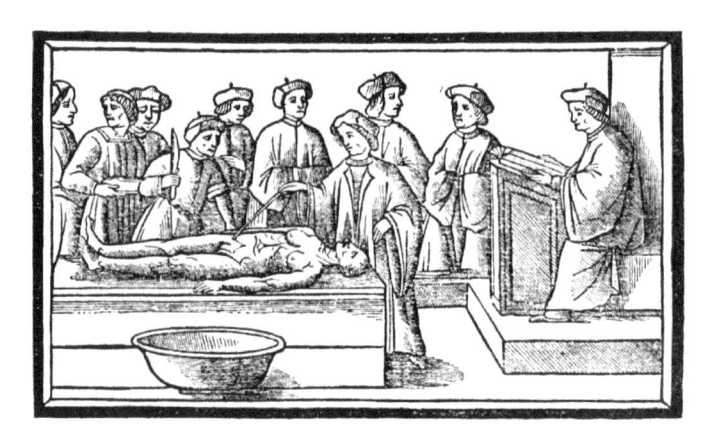

DISSECTION BY BERENGARIO, 1535

Giacomo Berengario, Jacobus Berengarius Carpensis, also known as Carpus, was born in the small town of Carpi, in the Duchy of Modena, in the year 1470. His father, who was a surgeon, directed his studies, and for a time he was placed under the instruction of the learned Aldus Manutius. Graduating in medicine from the University of Bologna, Berengario became noted for his skill in surgery and anatomy. He taught these branches in Pavia, and was a member of the Bologna faculty from 1502 to 1527. Then he practiced for a time in Rome, where he amassed a fortune by the treatment of the victims of syphilis. The last twenty years of his life were spent in Ferrara, where he died in 1550. Berengario was one of the restorers of anatomy. His first dissection is said to have been made in the house of Albert Pion, Seigneur de Carpi. This demonstration was given publicly upon the body of a pig. Soon the anatomist turned his attention to human subjects, of which it is said that more than a hundred passed beneath his scalpel.

Berengario's later years are said by Brambilla to have been made miserable by the machinations of the agents of the Inquisition, who objected to some of his opinions regarding the organs of generation. He was unjustly accused of dissecting living men—an accusation which arose from his statement that the surgeon should observe the anatomy of the living body whenever it was opened by wounds or accidents.

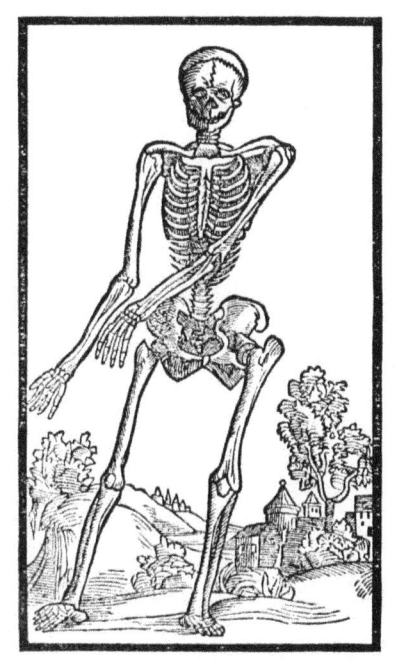

SKELETON BY BERENGARIO, 1523

Berengario determined to improve Mondino's book by making corrections in the text, and by adding suitable illustrations. No illustrations were to be found in the early editions of Mondino, and those which were added by later editors of the work were untrue to nature. To Berengario must be given the credit of furnishing some of the first anatomical illustrations that were published, and that were made from actual human dissections. These appeared in his "Commentaries of Carpus upon the Anatomy of Mundinus", (*Carpi Commentaria super Anatomia Mundini*), which was published at Bologna in 1521. The volume contains twenty-one plates which were cut in wood. They have been credited to the celebrated artist, Hugo da Carpi. While the drawing is somewhat coarse, the illustrations are true to nature and show a distinct advance over preceding pictures of this class. Berengario states that his plates will be of value not only to physicians and surgeons but also to artists (*et istae figurae etiam juvant pictores in lineandis membris*). Some of his figures are schematic; for example, those showing the abdominal muscles. So much better are his illustrations than those of his predecessors that it may fairly be claimed that Berengario was the first author to produce an illustrated anatomy.

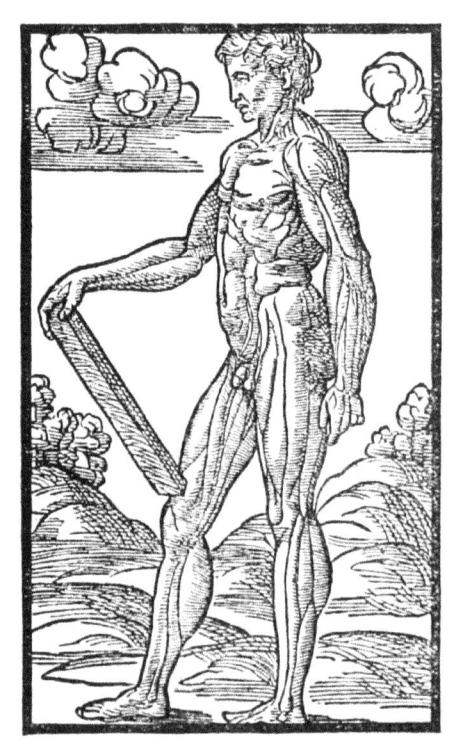

MUSCLES BY BERENGARIO, 1521

Berengario also wrote a "Short Introduction to the Anatomy of the Human Body", *Isagogae Breves in Anatomiam Humani Corporis*; and a work on Fracture of the Skull.

He was the first anatomist who described the basilar part of the occipital bone, the sphenoidal sinus and the tympanic membrane. Meryon[8] credits him with the "first correct description of the great omentum (gastrocolic) and transverse mesocolon; of the caecal appendix vermiformis, of the valvulae conniventes of the intestines; of the relative proportions of the thorax and pelvis in man and woman; of the flexor-brevis-pollicis; of the vesiculae seminales; of the separate cartilages of the larynx; of the membranous pellicle in front of the retina (attributed to Albinus); of the tricuspid valve, between the right auricle and ventricle of the heart; of the semilunar valves at the commencement of the pulmonary artery; of the inosculation between the epigastric and mammary arteries, and an imperfect account of the cochlea of the ear". He was the first of the mediaeval anatomists to deviate from the Galenic teaching in regard to the structure of the heart. He diplomatically states that in the human subject the foramina in the cardiac septum are seen only with great difficulty (*sed in homine cum maxima difficultate videnter*).

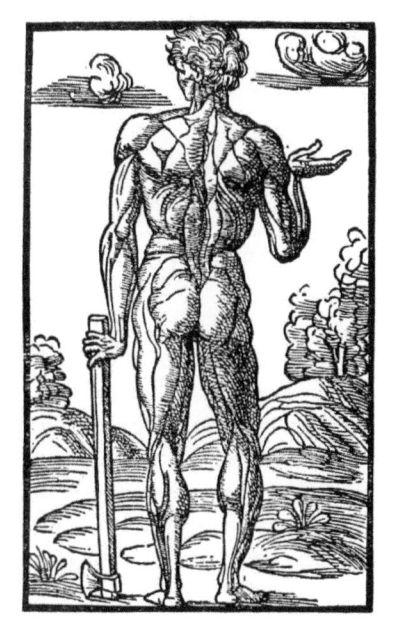

MUSCLES BY BERENGARIO, 1521

John Dryander

John Dryander, a German physician, whose true name was Eichmann, called himself Dryander in accordance with the custom of adopting names derived from the Latin or Greek languages. He was born about the year 1500 in the Wetterau in Hesse. After obtaining proficiency in mathematics and astronomy, he went to Paris where he studied medicine for several years. Returning to Germany, he engaged in the study of practical anatomy and became a professor in Marburg, in which city he died in the year 1560. He is said to have conducted the first dissections that were made in Marburg, where he taught anatomy for twenty-four years, or from 1536 to 1560.

DRYANDER

Dryander, although he was a partisan of Mondino and da Carpi, and was a fierce and sometimes an unfair opponent of Vesalius, deserves to be regarded as one of the restorers of anatomy. He made several observations upon the distinction between the cortical and the medullary portions of the brain; and was one of the earliest practical anatomists of the sixteenth century to furnish anatomical illustrations. He made important astronomical observations and was the inventor of several useful instruments. He was the author of three medical works of which two were upon anatomy. His *Anatomia Mundini*, which was published at Marburg in 1541, contains forty-six plates, many of which have been copied from Berengario's work.

ANATOMICAL FIGURE BY ESTIENNE, 1545

Charles Estienne

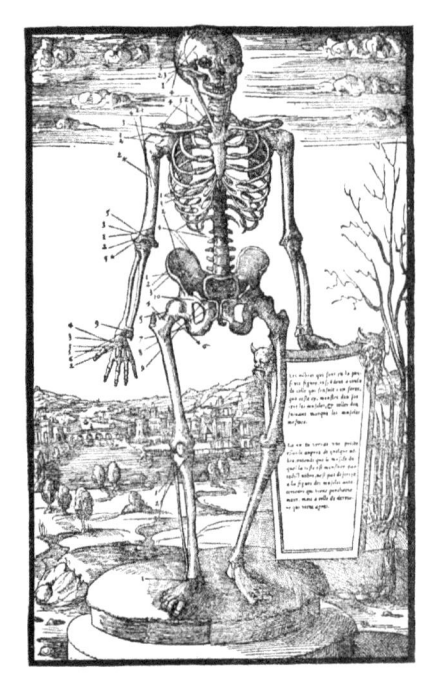

SKELETON BY ESTIENNE, 1545
(Reduced one-half)

Charles Estienne, better known by the name of *Carolus Stephanus*, was a French anatomist whose work is worthy of remembrance. Born in the early part of the sixteenth century, he was given an excellent education. He belonged to a noted Huguenot family of scholars and printers who have made the Estienne name famous. Robert Estienne, the brother of Charles, became the victim of religious persecution; he was obliged to flee to save his life, and for a time the publishing business was conducted by Charles Estienne. The latter also suffered for his faith; he was thrown into a dungeon, where he died in the year 1564. Charles Estienne wrote numerous books on literature, history, forestry and botany. His anatomical treatise, *De Dissectione Partium Corporis Humani*, appeared at Paris in 1545 with sixty-two full page plates which combine anatomical clearness, beauty of form, and artistic representation. A French translation of Estienne's Anatomy was published in 1546. This work was printed as far as the middle of the third book as early as the year 1539: some of the plates are dated as early as 1530. The illustrations have been excellently cut in wood; many of them show the entire body, with much ornamentation, so that the proper anatomical part seems small and irrelevant. Some of the plates show the subject in picturesque and even loathsome attitudes. The text of this work is especially valuable for the history of anatomical discovery. Although he was an ardent Galenist, Estienne made numerous original observations in anatomy. He described the synovial glands, a discovery which has been credited to Clopton Havers. Estienne was the first anatomist to discover the canal in the spinal cord; he described the capsule of the liver, a tissue which bears Glisson's name; and differentiated the eight pair from the sympathetic nerves. He was the first anatomist to see and describe the valves in the veins, which he called *apophyses venarum*— discovery which has been claimed for Jacobus Sylvius, Cannanus, Amatus and Fabricius.

The question of priority in the discovery of the valves of the veins gave rise to much controversy. It is reasonable to assume that these structures were noticed independently by all of the anatomists whose names are mentioned above.

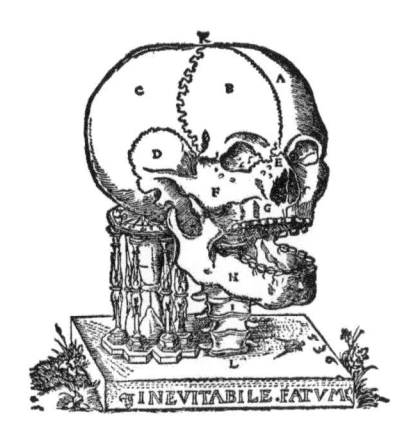

SKULL BY DRYANDER, 1541

CHAPTER FOURTH
Vesalius's Early Life

Andreas Vesalius, or Wesalius as the family name was inscribed prior to the year 1537, was born in Brussels on the last day of the year 1514. From astrological observations made by Jerome Cardan we learn that this event occurred about six o'clock in the morning, and under favorable stellar auspices. The placenta and caul, to which popular belief ascribed remarkable powers, were carefully preserved by the mother.

The Vesalius family originally was named Witing, (*Witting, Wytinck, Wytings*, according to various authorities) and adopted the name Wesalius from the town of Wesel, (*Wesele, Vesel*), in the Duchy of Cleves, which the family claimed as their native place. The three weasels (*Flemish*—"Wesel"), found in the Vesalian coat of arms, testify to this origin.

It may be said with truth that medical learning ran in the blood of the Vesalius family. Andreas's great-great-grandfather, Peter Wesalius, wrote a treatise on some of the works of Avicenna and at great cost restored the manuscripts of several medical authors. Peter's son, John Wesalius, held the responsible position of physician to Mary of Burgundy, the first wife of Maximilian the First; in his old age John taught medicine in the University of Louvain. From that time the Vesalius family was closely associated with the Austro-Burgundian dynasty. Eberhard, son of John Wesalius, served as physician to Mary of Burgundy; he died before attaining his thirty-sixth year, and was long survived by his father. Eberhard, who was the grandfather of Andreas, wrote commentaries upon the books of Rhazes and on the *Aphorisms* of Hippocrates. He was also noted as a mathematician. Eberhard's son Andreas, the father of the anatomist, was apothecary to Charles the Fifth and to Margaret of Austria. He accompanied the great Emperor upon his numerous journeys and military expeditions. In 1538 he presented Andreas's first anatomical plates to the Emperor, and thus opened the way to the court to his son. The father remained in the imperial service until the day of his death, which occurred in 1546. Andreas's mother, Isabella Crabbe, exercised a great influence upon the youth whom she believed to be destined to accomplish great things. She it was who preserved the manuscripts and books of the Vesalian ancestors. Isabella happily lived long enough to see the *Fabrica*, to witness the intellectual triumph of her son, and to know of his activity at the Spanish court.

THE OLD UNIVERSITY OF LOUVAIN
(Erected early in the Fourteenth Century. The New Building dates from 1680)

Little is known of the youth of Vesalius. The traditions of his ancestors, their accomplishments in the field of letters and in medicine, and their loyalty to their sovereigns, were themes which his mother must have recounted with pleasure. At an early age Andreas was sent to the neighboring city of Louvain, whose University, founded in the year 1424, in the early part of the sixteenth century eclipsed many institutions of greater age, and in the number of its students ranked second only to the University of Paris. The theologians of Louvain were noted for their orthodox Catholicism; from the very first days of religious controversy they had battled strongly against the rising tide of the Reformation. Her professors of jurisprudence and of philosophy were men of eminent talents. Within the University were four literary schools which were named *Paedagogium Castri*, *Porci*, *Lilii*, and *Falconis*, from their insignia:—a fort, a pig, a lily, and a falcon. Here also was the *Collegium trilingue Buslidianum*, which was founded by Hieronymus Busleiden (+1517) for teaching the Greek, Hebrew and Latin languages. Vesalius selected the *Paedagogium Castri* which he fondly mentions in laudatory terms in his *Fabrica*. Here, and in the Busleidinian College, he obtained that thorough knowledge of ancient languages which, in later years, astonished his hearers and served him well in numerous literary controversies. The names of Vesalius's teachers are unknown, although Adam[9] states that John Winter of Andernach was his professor of Greek. Vesalius speaks scornfully of one of his teachers, a theologian, who, in trying to explain Aristotle's *De Anima*, used a picture of the *Margarita Philosophica* to show the structure of the brain. Among Vesalius's school companions were Gisbertus Carbo, to whom the anatomist presented the first skeleton which he articulated (*Fabrica*, 1543, page 162); and the younger Granvella, who later was Chancellor to Charles the Fifth.

At an early age Vesalius possessed a desire to study the structure of the human body. His powers of observation were precociously developed. When a boy, learning to swim by the aid of bladders filled with air, he noted the elasticity of these organs, and he referred to the incident in his *Fabrica* (1543, page 518). When little more than a child, he tired of dialectics and tried to learn anatomy from the scholastic writings of Albertus Magnus and of Michael Scotus. He soon discovered that the true road to anatomical science led, not through books but through the actual handling of the dead tissues. He began the practical study of anatomy by dissecting the bodies of mice, moles, rats, dogs and cats.[10]

CHAPTER FIFTH
Sojourn in Paris

One thought was uppermost in the mind of Vesalius, and that was to follow the profession of his ancestors, just as in ancient Greece the sons of the Asclepiadae naturally adopted the vocation of their fathers. Andreas possessed an excellent preliminary education and was especially proficient in the Greek and Latin languages; he also knew something of Hebrew and much of Arabic. It was in the year 1533 that the young Belgian travelled to Paris for the purpose of obtaining a medical education. At that time the French capital was the Mecca of the medical world— Paris, that city where classical medicine first secured support (*ubi primum medicinam prospere renasci vidimus*)[11]. In Paris, under the leadership of Budaeus, Humanism had enjoyed a rapid growth; and here Petrus Brissotus, after gaining the doctor's cap in the year 1514, produced a revolution by delivering his lectures from the books of Galen in place of the treatises of Averröes and of Avicenna. At his own expense Brissotus published Leonicenus's translation of Galen's *Ars Curativa*, in order that his pupils might not be misled by the incorrect text of the Arab authors. It will be recalled that, long before this time, classical Greek and Latin medical literature had passed through the distorting crucible of Saracenic translations. At this period medical science, purified from Arabic dross, was taught in a splendid manner in Paris by such eminent professors as Jacobus Sylvius, Jean Fernel, and Winter of Andernach. At their feet sat young men from the remotest parts of Europe.

The most popular of the Paris teachers was Jacobus Sylvius, or Jacques Dubois, whose Latinized name is perpetuated in anatomical nomenclature. He was born at Louville, near Amiens, in 1478. In his early years he was noted for his scholarly attainments in the Greek, Latin and Hebrew languages and was the author of a French grammar. His anatomical knowledge was gained under Jean Tagault, a famous Parisian practitioner and surgical author.

SYLVIUS

Sylvius was noted for his industry, for his eloquence, and above all for his avarice. It was the inordinate desire for money which led him to abandon philology for medicine. While studying under Tagault he began a course of medical lectures, explanatory of the works of Hippocrates and Galen, with such success that the Faculty of the University of Paris protested on the score that Sylvius was not a graduate. He then went to Montpellier, whose medical professors had long held a high position, where, according to Astruc, he received the doctor's cap at the end of November, 1529. He was then above fifty years of age. Armed with this degree, he returned to Paris and immediately entered the lists as an independent medical teacher, but was again halted by the Faculty who ruled that he must first receive the Bachelor's degree. This he gained on June 28, 1531. Sylvius then resumed his lectures with such success that his classes in the Collége de Tréguier numbered from four to five hundred, while Fernel, who was a professor in the Collége de Cornouailles, lectured to almost empty benches. In 1550, Henry the Second named Sylvius Professor of Medicine, as the successor of Vidus Vidius, in the recently established Collége de France. Sylvius died January 13, 1555, and was interred in the paupers' cemetery as he had wished.

Sylvius was not only an eloquent lecturer but he was also a demonstrative teacher. He was the first professor in France who taught anatomy from the human cadaver. In his lectures on botany he used a collection of plants to elucidate the subject. His chief fault was a blind reverence for ancient authors. He regarded Galen's writings as gospel; if the cadaver presented structures unlike Galen's description, the fault was not in the book but in the dead body, or, perchance, human structure had changed since Galen's time! In one of his early books[12], Sylvius declared that Galen's anatomy was infallible; that Galen's treatise, *De Usu Partium*, was divine; and that further progress was impossible!

The character of Sylvius was contemptible. He was a man of vast learning and at the same time was rough, coarse and brutal. His avarice led him to endure the cold winters of Paris without the benefit of a fire; in severe weather he would play at football, or engage in other violent exercise in his room, to save the cost of fuel. Once, and once only, did his friends find him hilarious; they wondered and asked the cause. Sylvius said he was happy because he had dismissed his "three beasts, his mule, his cat and his maid". He was notoriously rigid in exacting his fees from students, and on one occasion he threatened to stop his lectures until two delinquents should pay their dues. Although he was supposed to have amassed great wealth, little of it was found after his death, and these sums were secreted in secluded places. In 1616, when his former residence in the *rue Saint-Jacques* was demolished, numerous gold pieces were found. His reputation for miserliness followed him beyond the grave, as witness his epitaph:

Sylbius hic situs est, gratis qui nil dedit unquàm,

Mortuus et gratis quod legis ista dolet.

"Sylvius lies here, who never gave anything for nothing:

Being dead, he even grieves that you read these lines for nothing."

In controversies he was violent and vindictive—a pastmaster in the use of bitter language. Jealous of the fame of other anatomists, he was particularly enraged when, in later years, he was opposed by Vesalius. Sylvius spoke of him not as Vesalius, but as *Vesanus*, a madman, who poisoned Europe by his impiety and clouded knowledge by his blunders. Such was the man who, in the mid-part of the sixteenth century, filled the position of highest honor in the Medical Faculty of the Collége de France[13].

Sylvius rendered valuable service in naming the muscles which, prior to his time, were designated by numbers. These, says Northcote[14] "were differently applied by almost every author; so that it was the description, and not the name, that must lead one to know what part was meant by such authors; and this required a previous thorough knowledge of anatomy". He is the first writer who mentions colored injections and is supposed to have discovered this useful adjunct of anatomical study. He was the first anatomist who published satisfactory descriptions of the pterygoid and clinoid processes of the sphenoid bone, and of the os unguis. He gave a good account of the sphenoidal sinus in the adult but denied its existence in the child, as had been affirmed by Fallopius[15]. Sylvius also wrote intelligently concerning the vertebrae but incorrectly described the sternum. His observation concerning the valves in the veins gave rise to much discussion; the honor of priority in the discovery, however, belongs to other anatomists—Estienne and Cannanus. His discoveries in cerebral anatomy have caused his name to be attached to the *aqueduct*, the *fissure* and the *artery of Sylvius*.

The manner in which Sylvius conducted his anatomical course is known to us by his own writings, by the testimony of Moreau[16], and by that of Vesalius[17]. Thus the course for the year 1535 began with the reading, by Sylvius, of Galen's treatise *De Usu Partium*. When the middle of the first book was reached, Sylvius remarked that the subject was too difficult for his students to understand and that he would not plague his class with it. He then jumped to the fourth book, read all to the tenth book, discussed a part of the tenth and omitting the eleventh, twelfth and thirteenth, he took up the fourteenth and the remaining three books. Thus he omitted all that Galen had said concerning the extremities. A second Galenic work which Sylvius used was the anatomico-physiologic treatise, *De Musculorum Motu*. Not infrequently the professor was unable to demonstrate in dissection the parts on which he had lectured. Thus, on one occasion, the students succeeded in finding the pulmonary and aortic valves which Sylvius had failed to find on the preceding day.

Joannes Guinterius of Andernach

Another famous member of the Paris Faculty of this period, and a man whose life-story reads like a romance, was Joannes Guinterius, the beggar of Deventer. Guinterius (Gonthier, Guinther, Guinter, Winter, or Winther), who is often called John Winter of Andernach, from the name of the town in which he was born, lived between the years 1487-1574, and rose to eminence in both the literary and the medical worlds. Born of humble parents, he was sent at an early age to the University of

Utrecht. Leaving this institution because of his poverty, he went to Deventer where he was reduced to the necessity of begging in the streets. He drifted to the University of Marburg, and here displayed such brilliant talents that he soon obtained employment as a teacher in the small town of Goslar, in Brunswick. His growing reputation for learning led to his appointment to the chair of Greek in the noted University of Louvain.

WINTER OF ANDERNACH

Desiring to study medicine, Guinterius went to Paris in 1525; he received the Bachelor's degree in 1528, and the full medical title two years later. He passed a brilliant examination which won for him the commendation of the most eminent professors. Remaining in Paris, he engaged in practice and in teaching, and rapidly rose to eminence. In addition to conducting courses in anatomy, he translated into Latin the writings of the most noted Greek medical authors of antiquity—the books of Galen, of Oribasius, of Paul of Aegina, of Caelius Aurelianus, and of Alexander of Tralles—all of which were held in high esteem in the sixteenth century. His fame reached far beyond the boundaries of France. Christian III., the enlightened king of Denmark, who was noted for his love of literature, sought to attach him to the Danish court, but the honor was refused. Having become a convert to the religious views of Luther, Guinterius found that his life was in danger; he left Paris and resided for a time in Metz. He soon removed to Strassburg, where he was received with distinguished honors and was appointed to a professorship in the University. Owing to the activity of his enemies, his position became insecure; accordingly, he resigned his chair and spent a considerable time in travelling throughout Germany and Italy. In the year 1562, Ferdinand I., in appreciation of the great merits of Guinterius, raised him to the highest distinction by placing him among the nobles of the land; and thus the beggar of Deventer became a nobleman of Strassburg. His life ended October 4, 1574.

Like Sylvius, Guinterius was a teacher of men who became greater than himself—Vesalius, Servetus and Rondelet sat upon his benches. Like Sylvius, he placed his faith in Galen and failed to grasp the great truth that anatomical science is based, not on the writings of the Fathers but on dissection of the dead body.

Jean Fernel

JEAN FERNEL

The third bright star of the Paris constellation was Jean Fernel (1485-1558), of Amiens, who was regarded as the ablest physiologist of his time and was physician-in-ordinary to Henry the Second. Fernel dipped deeply into philosophy, geometry and mathematics. Before entering the medical profession he issued three books on mathematic and geometric subjects. He received the medical degree in 1530, but continued his study of mathematics with such ardor that he was almost ruined financially. On the advice of his friends he entered upon the practice of medicine in Paris and met with remarkable success. He was skilled in anatomy and surgery and accompanied his sovereign upon numerous military expeditions. His medical writings are contained in many volumes and concern a variety of subjects, such as physiology, therapeutics, surgery, pathology, the treatment of fevers and the venereal diseases.

Fernel's medical views were powerfully influenced by the teachings of an unfortunate French philosopher, Pierre de la Rameé, or Ramus, who, like many other Protestants, lost his life on Saint Bartholomew's Night. Brutally assassinated, his body was dragged through the streets of Paris and then was thrown into the Seine; but his system of philosophy survived and exercised a potent influence until it was eclipsed by the doctrines of Descartes.

Ramus, who was an uncompromising opponent of the Aristotelian philosophy, pointed out the defects and suggested the reforms in the system of University education. He compared the teaching of medicine with that of theology, much to the disparagement of the latter:—"The reason", said he, "why medicine is better taught, and the lectures are better attended than in theology is, that those who teach it know it, and practice it, and their disputations are chiefly on the books of Hippocrates and Galen; whilst the theologians observe a strict reticence on questions of the Old Testament, which they read in Hebrew, as well as of the New, which they read in Greek, but display their learning in subtle questions respecting the pagan philosophy of Plato and Aristotle".[18] Ramus endeavored to withdraw the minds of both physicians and medical students from the authoritative dogmas of the ancient physicians and to substitute therefor the intelligent study of Nature. The practical trend of his mind is shown in his suggestion that institutions should be arranged for clinical teaching.

RAMUS

Just as Ramus had become an Eclectic in philosophy, so Fernel sought the best from various sources and different medical systems. Like Ramus, he cast off the yoke which authority had placed upon him; and proposed carefully planned principles which should lead to the discovery of truth. Like Ramus, Fernel presented his views in a clear style and in better order than was to be found in the writings of his predecessors. Like Ramus, he adopted the good and rejected the bad, regardless of whether it had been said by Aristotle, or by Galen, or by Hippocrates. Fernel was a reformer who stood for freedom of thought, which, up to his time, had suffered from the despotism of the scholastics. Although many of Fernel's physiologic and pathologic ideas seem ridiculous when viewed in the light of modern knowledge, yet he deserves praise for daring to oppose ancient dogmas, and for pointing the road to progress. In breadth of view, Fernel was far superior to Sylvius and Guinterius.

The anatomical teaching in Paris in the early part of the sixteenth century was far from satisfactory. There was too much lecturing and theorizing from Galen's texts, and too little of actual dissection. Vesalius, who was not backward in his criticisms, says that the dissections were made by ignorant barbers, and during the whole time that he was in Paris he never saw Guinterius use a knife upon a cadaver. Only at rare intervals was a human body brought into the amphitheatre, and then the dissection lasted less than three days. It comprised only a superficial study of the intestines and abdominal muscles; no other muscles were studied. The bones, veins, arteries and nerves were almost wholly ignored. The great lights of the Paris profession were totally unfit to give to the young Belgian what was his heart's desire. They were ignorant and knew it not. It is not surprising that, on more than one occasion, Vesalius brushed the ignorant prosectors aside, took the knife into his own hands, and carried out the dissection in a systematic manner. His zeal and learning won the admiration of Guinterius who spoke of Vesalius and Servetus in loving terms;—"first Andreas Vesalius, a young man, by Hercules! of singular zeal in the study of anatomy; and second, Michael Villanovanus (Servetus), deeply imbued with learning of every kind, and behind none in his knowledge of the Galenic doctrine. With the aid of these two, I have examined the muscles, veins, arteries and nerves of the whole body, and demonstrated them to the students".[19]

Vesalius must have had many blue days in Paris—days when he longed to have a free hand in dissection. A weaker character than his would have fitted peacefully into the established order of things, but not of such stuff was Andreas made. The difficulties which beset his path only stimulated him to work the harder; he firmly resolved to devote his energy, his talents and his life to anatomical study and teaching. He decided to secure the opportunity to dissect the human body and to rival the ancient Alexandrian professors who taught the subject. "Never", he says, "would I have been able to accomplish my purpose in Paris, if I had not taken the work into my own hands". The Book of Nature which Sylvius lauded, but kept his pupils from studying, was now opened by Vesalius. He dissected numerous dogs and studied the only part of human anatomy that was available, namely, the bones. In his search for materials for a skeleton he haunted the Cemetery of the Innocents. On one occasion, when he went to Montfauçon, the place where the bodies of executed criminals were deposited and bones were plentiful, Vesalius and his fellow-student were attacked by fierce dogs. For a time the young anatomist was in danger of leaving his own bones to the hungry scavengers. By such dangers he gained what the Paris professors could not supply. He became a master of the osseous system, so much so that, when blindfolded, he was able to name and describe any part of the skeleton which was placed in his hands. His talents were recognized by both professors and students; and at the third anatomy which he attended in Paris he was requested to take charge of the dissection. To the satisfaction of the students, as well as to the astonishment of the barbers, he made an elaborate dissection of the abdominal organs and of the muscles of the arm.

VIVISECTION OF A PIG
(From the "Fabrica", 1543)

CHAPTER SIXTH
Vesalius Returns to Louvain

In the latter part of the year 1536, owing to the outbreak of the third Franco-German war, Vesalius returned to the University of Louvain. During this period he secured a human skeleton by secret means. Accompanied by his faithful friend, Regnier Gemma, known as a mathematician as well as a physician, Vesalius visited the gallows outside the walls of Louvain in order to search for bones. Here he found a skeleton which was held together simply by the ligaments and still possessed the origins and insertions of the muscles. Morley states that the body was that of "a noted robber, who, since he deserved more than ordinary hanging, had been chained to the top of a high stake and roasted alive. He had been roasted by a slow fire made of straw, that was kept burning at some distance below his feet. In that way there had been a dish cooked for the fowls of heaven, which was regarded by them as a special dainty. The sweet flesh of the delicately roasted thief they had preferred to any other; his bones, therefore, had been elaborately picked and there was left suspended on the stake a skeleton dissected out and cleaned by many beaks with rare precision. The dazzling skeleton, complete and clean, was lifted up on high before the eyes of the anatomist, who had been striving hitherto to piece together such a thing out of the bones of many people, gathered as occasion offered".

Such a prize could not be lost. With Gemma's assistance Vesalius climbed the gallows and secured the skeleton which he secretly conveyed to his home. The treasure, however, was not complete; one finger, a patella and a foot were missing. To this extent was Vesalius the owner of a human skeleton. In supplying the missing parts Vesalius was obliged to incur new dangers. He stole out of the city in the nighttime, climbed the gallows unaided, searched through the mass of decaying bodies, and, having found the coveted bones, he stole into the city by another gate. These secret expeditions, however, soon became unnecessary, for the Burgomaster of Louvain generously furnished an abundance of material for Vesalius's students.

It was at this period—late in the year 1536 or early in 1537—that Vesalius conducted the first public anatomy that had been held in Louvain in eighteen years. He performed the dissection and lectured at the same time, which was an innovation. Some remarks he made concerning the seat of the soul caused him to be critised by the theologians. A further cause for suspicion was his association with such firm Protestants as Guinterius and Sturm of Paris; and his friendly relations with the publisher Rescius, and the physician Velsius. Fortunately the suspicion of heresy did not lead to any formal charges, but the affair seems to have rankled in his memory and some years later, in his *Fabrica*, he sought to clear his name of even the appearance of heresy.

Vesalius began his career as an author by issuing a paraphrase, or free translation, of the ninth book of the *Almansor* of the celebrated Rhazes[20]. This book, *liber ad Almansorem*, or work dedicated to the Caliph Al-Mansûr, was written by a learned Arab physician who lived between the years 860-932. The *Almansor* consists of ten books and was designed by the author for a complete body or compendium of Physic. The first book treats of anatomy and physiology; the second, of temperaments; the third, of food and simple medicines; the fourth, of means for preserving health; the fifth, of skin diseases and cosmetics; the sixth, of diet; the seventh, of surgery; the eighth, of poisons; the ninth, of treatment of all parts of the body; the tenth, or last book, deals with the treatment of fevers. The ninth book, which Vesalius translated from the barbarous version into a readable form, was so highly prized in mediaeval times that it was read publicly in the schools and was commentated by learned professors for more than a hundred years. By this publication Vesalius furnished a valuable contribution to medical literature. The numerous marginal and interlinear notes, which he supplied, show his intimate acquaintance with classical literature as well as with materia medica. Vesalius emphasizes the fact that the book of Rhazes contains many remedies which were unknown to the Greeks. The value of his edition was increased by the presence of original drawings of the plants mentioned in the text.

CHAPTER SEVENTH
Professor of Anatomy in Padua

Shortly after the publication of his *Paraphrasis in nonum librum Rhazae*, Vesalius journeyed into Italy. It was in the year 1537 that he entered the prosperous and enlightened city of Venice. Here the study of anatomy not only was not tabooed, but was encouraged, particularly by the Theatin monks who devoted themselves to the care of the sick. At the head of this order stood two remarkable men: J. Peter Caraffa, who later ascended the papal throne as Paul IV.; and Ignatius Loyola, the founder of the Jesuits. It is a strange circumstance that two strong characters so dissimilar as were Vesalius and Loyola should meet as co-workers in the same field. The one was filled with a thirst for anatomical knowledge, and was dreaming of the day when his *opus magnum* should revolutionize an important science; the other was enthused with visions of the world-wide acceptance of the doctrines of Catholicism. They met again, in 1543—the year which marks two important events, namely, the publication of the *Fabrica*, and the full recognition of the Jesuits by the Pope.

In Venice the young anatomist entered into various lines of activity. He experimented with a new remedy, the China root, and besought his acquaintances to observe its effects in cases of pleurisy. He solicited anatomical material and possibly may have conducted a public demonstration in anatomy, although this is uncertain. He practiced minor surgery; he leeched and opened veins, particularly the popliteal vein which the barbers of that day did not venture to touch. In Venice he fortunately met his countryman, Jan Stephan van Calcar, who was soon to furnish the drawings for Vesalius's first anatomical plates.

ANATOMICORVM INSTRVMEN-
TORVM DELINEATIO.

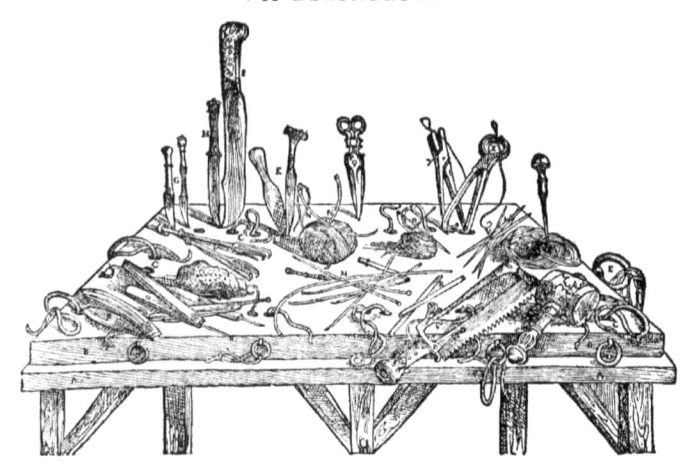

INSTRUMENTS USED IN DISSECTION
(From the "Fabrica", 1543)

In order to gain all the rights and privileges of a full-fledged physician, Vesalius settled in Padua. On the 6th day of December, 1537, shortly after having received his degree as Doctor of Medicine, Andreas Vesalius of Brussels was appointed Professor of Surgery with the right to teach Anatomy in the famous University of Padua. This, says Fisher, "was the first purely anatomical chair ever instituted".

From his own writings and from the manuscript notes of his loyal student, Vitus Tritonius, a fairly good idea of Vesalius's teaching can be given. The first act of the young Paduan professor was to improve the course in anatomy. Here, as he had done previously at Louvain, Vesalius discharged the entire duties of the professorship. He acted as lecturer, demonstrator and dissector. Dissatisfied with the ignorant barbers, he ignored them and employed his students as assistants. He resorted to all possible means to obtain anatomical material, much of which was secured by stealth.

The aula in which Vesalius conducted his course was built of wood and was capable of holding five hundred persons. In the centre of the room was a table under which was a receptacle containing bones and joints. An articulated skeleton was placed in an upright position at one end of the table. In this elegantly appointed room, before an audience of distinguished laymen and students, the instruction in anatomy was given. The course was a strenuous one, occupying practically the entire day for a period of three weeks, and comprising not only human but also much comparative anatomy. The vivisection of dogs, pigs, and rarely of cats, was a regular part of the course. Drawings were used to elucidate the relations between the skeleton and the soft parts; and frequently Vesalius marked the outlines of the joints upon the skin of the subject. He also marked the cranial sutures with ink. His anatomical charts were the work of his own hand; at times he drew the pictures in the presence of his audience. His dissections were made with extreme neatness and dexterity. He used but few instruments and these were of the simplest kind: knives of different shapes, hooks, cannula, catheter, sounds, bristles, hammer, saw, needles, thread and a sponge. Forceps and injection apparatus were not used; he rarely used scissors. Much of the actual separation of tissues was done by the aid of the finger-nails. A vivisection board completed the list *de instrumentis quae anatomes studioso debent esse ad manum.*

Let us now follow one Vesalius's public courses in anatomy. It is the month of December, in the year 1537. The report has spread that the young Belgian professor will begin his course. Long before the hour set for the lecture, every available seat has been taken and many persons are standing. An audience comprising the professors of the University, the students of medicine, officials of the city of Padua, and learned persons of all ranks, including members of the clergy, numbering more than five hundred persons, has assembled to do honor to the professor of anatomy.

Vesalius comes into the arena and walks to the table which is closely surrounded by his auditors. He wastes no time; after a few preliminary remarks on the importance of anatomy and the methods of acquiring a knowledge of this science, he launches into the practical demonstration. After rapidly pointing out the divisions of the body, and demonstrating the skin, joints, cartilages, ligaments, glands, fat and muscles, he passes to the more complex parts, all of which are shown upon the skinned body of a dog or of a lamb, in order to conserve the human material. Now the human cadaver is placed on the table; all eyes are turned upon it, for such a demonstration occurs only at long intervals. Vesalius speaks first of the difference in the structure of joints at different ages and in different sexes, illustrating his remarks by means of drawings and by an abundant supply of bones of man and of the lower animals.

Now comes the dissection. This is made rapidly and in regular order. Its course depends upon the amount of material at hand; if the professor resorts to two bodies, as in the year 1538, the demonstration is handled in grand style. Vesalius uses the first body for a comprehensive examination of the muscles, ligaments and viscera; whilst the second cadaver is devoted to the relations of the veins, arteries, nerves and viscera. The text of the *Fabrica* is written according to this plan of public dissection.

At times Vesalius attempted to teach the whole of anatomy on one cadaver. In this event, osteology was followed by the dissection of the abdominal muscles layer by layer, the demonstration closing with an examination of the entire contents of the abdomen. The pelvic organs were reached by incision and separation of the symphysis pubis. If the cadaver was that of a female, the dissection began with the mammary glands and then passed to the inferior venter. In pregnancy the foetal membranes were removed intact, and were placed in a vessel filled with water. The foetus was opened and its anastomosing vessels were found. For demonstrating the cotyledons, the uterus of a sheep or goat was used. After the thorax had been raised by means of a log or brick, Vesalius passed to the face and the anterior part of the neck, freely exposing the muscles on one side and the vessels and nerves on the other. Then followed the unilateral preparation of the muscles of the shoulder and back, then those of the mouth, which were approached by means of division of the lower jaw; and, finally, the pharynx and the larynx were exposed. The rectus anticus muscle was next brought into view, whereupon Vesalius detached the head from the vertebral column. Decapitation was followed by an examination of the cranium; the skull-cap was sawed and the brain was dissected in its natural position. Then came the examination of the eye, which Vesalius dissected in two ways: either by a complete section, or layer by layer from without inwards.

The ear and the cavities of the frontal and sphenoidal bones were next opened, provided these bones were not needed for the setting up of a skeleton. Finally he took up the extremities, demonstrating the muscles of an arm and a leg on one side, and the nerves and vessels on the other. The anatomy lesson ended with the introduction of numerous vivisections.

Vesalius could not entirely escape disputations, but he gave to them a close anatomic basis. Theoretical physiology was repugnant to him; for him physiology was not speculation but the sequel of anatomic research. If he at times gave free reign to his views, he indicated them as mere theories. He did not ignore pathologic conditions, but he handled them as briefly as possible. Fearing to tire his audience with too much variety, he confined his students closely to the structure of the human body.

The merit of Vesalius's public dissections, and the impression which they made upon his auditors, can be appreciated only by comparison with similar demonstrations made by his predecessors. The large and enlightened audience remained day by day for a period of three or four weeks. He says not a word about the physical and mental strain incident to such a strenuous course, in which his entire time was employed. The courses brought great financial profit to the professor.

On two occasions, probably in the years 1539 and 1540, Vesalius was called from Padua to Bologna to conduct public dissections. This was a great honor, for Bologna was the city in which Mondino had revived the practical teaching of anatomy. These courses were conducted by Vesalius in a wooden building erected for that particular purpose. Here, as in Padua, the professor acted as demonstrator and lecturer, remaining in this ancient city for a period of several weeks. On the first occasion he was supplied with three human bodies and was enabled to handle the subject in grand style. At the first séance he engaged with the celebrated Professor Matthaeus Curtius, whose acquaintance he had made in 1538 while on a vacation trip, in a deep study of the question of venesection. Before a large and select assembly he demonstrated in all three bodies that Galen's description of the vena azygos was incorrect. On the second convocation Vesalius seems to have disposed of more bodies. He reviewed Galen's work on the joints, and by numerous specimens, which were prepared by the students, he demonstrated the difference in the ancient knowledge of the skeleton. On this occasion he undertook the complete dissection of an ape and presented its skeleton, as well as that of a man, to Professor John Andreas Albius, who held the chair of Hippocratic medicine in Bologna.

Little is known of the way in which Vesalius taught surgery. The first year he was in Padua, he began with Avicenna's treatise on tumors. According to the fragmentary notes in the college book of his ardent pupil, Vitus Tritonius, Vesalius compared

Avicenna's teachings with the classical works of Hippocrates, Galen, Paul of Aegina, and Aetius, explaining and correcting them.

INITIAL LETTER BY VESALIUS
(From the "Fabrica", 1543)

CHAPTER EIGHTH
First Contribution to Anatomy

Like all great teachers, Vesalius was ever mindful of the interests of his students. Soon after accepting the chair of Anatomy in Padua, he articulated a human skeleton for use in his class room. His next work was the preparation of a set of anatomical plates, *Tabulae Anatomicae*, which were intended to pave the way to anatomy for beginners. For the further benefit of his class, he edited an edition of Guinterius's *Institutionuin Anatomicarum*, which was issued in April, 1538.

Tabulae Anatomicae

The *Tabulae Anatomicae* were in the form of *Fliegende Blätter*, or loose leaves, and consisted of six plates which are now among the rarest of medical works. They bore the following title:

Tabulae Anatomicae. Imprimebat Venetiis B(ernardinus).
Vitalis Venetus sumptibus Joannis Stephani
Calcarensis Prostrant verò in officina
D. Bernardini. a. 1538.

In the preface Vesalius says that no one can learn either botany or anatomy from figures alone, but illustrations are a valuable means toward the imparting of knowledge. In publishing these plates he hopes to benefit those persons who had attended his public dissections. Not a line in these pictures is unnatural; all has been reproduced just as he had shown in his demonstrations. He gives due credit to van Calcar, the artist who made the drawings of the three skeletons. The other pictures were made by the author himself.

The *Tabulae Anatomicae* were arranged in the following order:—

I.—The Portal System and the Organs of Generation;

II.—The Venae Cavae and Chief Veins;

III.—The Great Artery—Arteria Magna—and the Heart;

IV.—The Skeleton in its Anterior View;

V.—The Skeleton in its Side View;

VI.—The Skeleton in its Posterior View.

The plates are of large dimensions, measuring over sixteen inches in length, and were cut in wood. Like those in the *Fabrica*, they were made in Italy. Owing to their transient use by medical students, the *Tabulae* were soon destroyed, although unauthorized editions were printed in several cities. The book was dedicated to Narcissus of Parthenope (Narciso Verdunno, or Vertuneo) who, in 1520, was first physician to the crown of Naples, and later, in 1524, was physician and councilor to Charles the Fifth. It is noteworthy that three of these plates deal with the skeleton, a subject to which Vesalius had given much attention. The absence of a plate showing the nervous system is also to be noted. Vesalius had such a plate prepared, and it appeared in a pirated edition of the *Tabulae* which was published at Cologne in 1539. The large size of these plates, their fidelity to nature, and the skill with which they were cut in wood, were features which showed to the world that a real master of anatomy had been born. The original drawings were made by Jan Stephan van Calcar, who probably also was the engraver.

Only two copies of the *Tabulae Anatomicae* are known. A fine edition of these plates, reproduced by photography, was privately issued in 1874 by Sir William Stirling-Maxwell, the talented author of the *Annals of the Artists of Spain*.

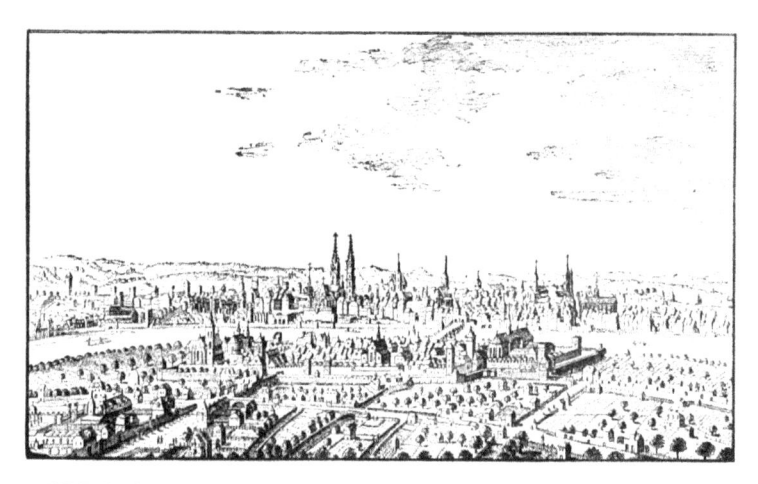

VIEW OF THE CITY OF BASEL IN THE SIXTEENTH CENTURY

CHAPTER NINTH
Publication of the Fabrica

On the first day of August, 1542, after three years of strenuous labor, Vesalius completed the *Fabrica*, and twelve days later he wrote the last word of the *Epitome*. The blocks for the *Fabrica*, and also those for the *Epitome*, were made in Italy. In the summer of 1542 they were conveyed to Basel by a merchant named Danoni and were safely delivered to the printer, Oporinus. They were accompanied by a long Latin letter, written by Vesalius to his friend, "Joannes Oporinus, professor of Greek letters in Basel". He begs Oporinus to take the greatest care that the printed illustrations shall correspond with the proofs which accompany the blocks. "Every detail must be distinctly visible, so that each cut shall have the effect of a picture". Early in the following year Vesalius went to Basel to superintend the printing of his books. While there, he conducted a demonstration in anatomy—the first which had occurred in that city since 1531—and presented the articulated skeleton of the subject to the University. Part of this skeleton exists today. It is thought to be the oldest anatomical preparation in existence.

The Fabrica

The heart of Vesalius must have filled with joy when he saw the final page of his book turned from the press. The treatise which founded modern anatomy bears this title:—

> *Andreae Desalii Brurellensis, Scholae medicorum Patabinae professoris, de humani corporis fabrica Libri septem. Basileae.*
> MDXLIII

A fortune was lavished upon the illustration and publication of this grand work. To use the words of Fisher, "it was and is a glorious book, a rare and precious monument of genius, industry and liberality". It abounds with curious initial letters bearing quaint and interesting anatomical conceits, each one teaching its lesson. One of these, reduced in size, introduces the present chapter; and it was this letter that Vesalius used in his opening sentence: *Os caeterarum hominis partium est durissimum & ardissimum, maximaque terrestre & frigidum, & sensus denique praeter solos dentes expers.*

JOANNES OPORINUS

The first edition of the *Fabrica* is a folio volume with magnificent illustrations on wood, all carefully printed by Joannes Oporinus (1507-1568) of Basel.

The title-page is a beautiful engraving which represents Vesalius at work dissecting a female subject. He is surrounded by interested spectators who crowd the amphitheatre. The abdomen of the subject is opened. Vesalius has raised his left hand; his right hand grasps a small rod which rests on the viscera. The great teacher is talking to his pupils. Placed at the head of the dissecting table is an upright skeleton which grasps a long staff with its right hand. In the audience are many persons of different rank. To the left a naked man is climbing a pillar, while to the right, and below, a dog is being brought into the arena. To the left, and below, is a monkey which appears to enjoy the demonstration. Above, in the architecture, we see the monogram of the publisher, Oporinus; in the centre, on a shield, are the three weasels of the Vesalius family, and below, is a shield which bears the privilegium. This old engraving is one of the most spirited and elaborate to be found in the whole range of medical literature. In the 1725 edition, for which Jan Wandelaar made copperplate reproductions of the original figures, the title-page is altered:—the monogram of Oporinus is absent and the architecture is slightly changed.

MARK OF OPORINUS

Who was the unnamed artist? It is noteworthy that Vesalius does not state who drew the illustrations, or who cut them in wood, for his *Fabrica*. He only states that this book has cost him a monstrous amount of labor in the preparation of the dissections, and in the directing of the eye, the hand, and the intelligence of the artist. He complains bitterly of the obstinacy of the artist, who, at times so tormented him that he—Vesalius—considered himself more unfortunate than the criminal whose body had been dissected[21]. It was probably owing to this unpleasant experience that Vesalius omitted the artist's name. The great anatomist speaks regretfully of the large sums which he was obliged to pay, in order to induce skilled artists to undertake this class of work. He states that they were much more interested in painting Venus and The Graces than in drawing pictures of skinned and foul smelling bodies. Moehsen[22] assumes that Vesalius had Titian in mind when he penned these thoughts, but this is questionable. It is not surprising that eminent artists should have disliked anatomical drawing, at a time when antiseptic injections and preserving fluids were not known. Foul odors had no terrors for the great Belgian, who haunted cemeteries for anatomical material and often kept parts of cadavers in his bedchamber for weeks at a time.

For a period of two centuries the Vesalian pictures were ascribed to Titian, but on insufficient grounds. The famous Venetian painter was over sixty years of age at the time of the publication of the *Fabrica*; his services were much in demand, and he was signally honored by the Spanish emperor, Charles the Fifth. His powers remained undiminished until shortly before his death, which occurred in 1576. He had the ability to make the Vesalian illustrations, but it is doubtful if he had the time. Although Titian may have taken an interest in these anatomical plates, it is not now believed that he drew them.

JAN STEPHAN VAN CALCAR

The Vesalian pictures have been attributed to Christoforo Coriolano; but he could not have been the artist, since his earliest work dates from 1568. He is known to have furnished the drawings for Jerome Mercurialis's *De Arte Gymnastica*, and for Vasari's *Lives of the Painters*. Roth is of the opinion that Vesalius himself made most of the illustrations; but such a view would credit the comparatively short and busy life of the great anatomist with too much accomplishment.

I conclude that the illustrations for the *Fabrica*, like the osseous figures in the *Tabulae Anatomicae*, which Vesalius issued in 1538, were made by Jan Stephan van Calcar (+1546), the favorite pupil of Titian. Sandrart[23] states that van Calcar made the drawings for the *Fabrica*; that he went to Venice in 1536 or 1537; that he studied under Titian; and that his paintings were of such merit that they were often mistaken for those of Titian, Raphael, and Rubens.

Van Calcar was a Fleming, a native of Kalcker in the Duchy of Cleves. The date of his birth is not known. His death occurred at Naples in 1546. He was highly esteemed by Vesalius who speaks of him as ranking "with the divine and happy wits of Italy". The anatomical plates which Vesalius issued in 1538 were made, he states, by van Calcar:—*sumptibus Joannis Stephani Calcarensis*. These plates, which appeared in the form of pictorial broad sheets, or *Fliegende Blätter*, may be likened to the Herald who goes in advance to announce the coming of the King. They were engraved on wood, and, like their companion pictures in the *Fabrica*, they were unprecedented in magnitude and in minuteness.

SECOND VESALIAN PLATE OF THE MUSCLES
(From the "Fabrica", 1543. Reduced one-half)

The Vesalian plates vary greatly in merit. The most satisfactory ones are those depicting the undissected body and the bones and muscles. The artist was not at his best in drawing the nervous system, although it is claimed that Vesalius had prepared his neurologic specimens with great care. For the use of artists, the best plates are the three skeletons and the four entire myologic figures in the *Fabrica*. The first myologic figure, showing a man who has been divested of all skin, fat, and superficial fascia, presents the muscles of the anterior portion of the body beautifully delineated. Vesalius took much pride in this plate, and directed the attention of artists to it. The second plate, which is constructed along similar lines, shows the body in its lateral aspect. The head is thrown slightly backward, the right hand pointing to the earth and the left raised towards the horizon, and the whole attitude of the subject calls to mind the position which an orator would assume when addressing an audience. The third myologic plate is similar to the first one, but the muscles of the face are exhibited to better advantage and the aponeuroses, absent in the first plate, are here present. The fourth plate, which is the ninth in Vesalius's work (*nona musculorum tabula*), presents the muscles of the posterior part of the body. The other myologic figures show the deeper muscles, layer by layer, and are of value to an artist who wishes to study the effect of their action upon the superficial parts of the body. Hence many of these figures have been reproduced in works on art-anatomy. The artist who studies these plates should remember that the figures in question are divested of skin, fat, and superficial veins—all of which must be supplied, in order to avoid giving too great prominence to the muscles. The two naked figures contained in the *Epitome* are properly clothed in skin and are of great artistic merit. They also are to be seen in numerous works on art-anatomy. Thus, in one of the earliest books on anatomy for the use of artists (*Abrégé d'anatomie accommodé aux arts de peinture et de sculpture.* Paris, 1667, 1668), Rogers de Piles and François Tortebat have used the three skeletons and seven myologic figures taken from the *Fabrica* and the *Epitome*. In the preface of his book, de Piles says that he does not think it is possible to produce better figures than those found in the works of Vesalius. That he was not alone in this opinion is shown by the fact that many other artists, who have composed treatises on art-anatomy, have drawn freely from the Vesalian storehouse. An Italian, Giacomo Moro, in his anatomy for the use of artists, (*Anatomia ridotta ad uso de' pittori e scultore.* Venice 1679), reproduced nineteen of Vesalius's figures in copperplate.

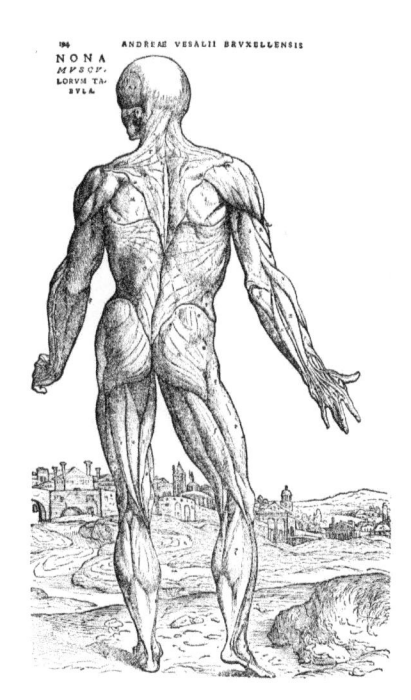

NINTH VESALIAN PLATE, OF THE MUSCLES
(From the "Fabrica", 1543. Reduced one-half)

The popularity of Vesalius's anatomical figures among painters was due, not only to the intrinsic worth of these illustrations, but also to the erroneous belief that the original drawings were the work of Titian. This opinion found expression on the title-pages of several works on art-anatomy. For example, in 1706, Moschenbauer, of Augsburg, issued a folio volume illustrated with Vesalian figures cut in wood, with this title:—*Andreae Vesalii, Bruxellensis, des ersten besten Anatomici, Zergliederung des menschlichen Körpers auf Mahlerey, and Bildhauer-Kunst gerichtet, die Figuren von Titian gezeichnet.* An anonymous book, *Notomia di Titanio*, appeared in Italy about the year 1670.

The Vesalian figures of the skeleton were also issued in single sheets with moralistic verses appended. Moehsen cites one of these with the inscription printed in French:

"De cet objet affreux tu parois rebutté,

Est c'est ce que dans peu cependant tu dois étre:

Apprens, mortel, a te connoître

Ce miroir est le seul, ou tu n'est point flatté".

Another legend reminds the reader that he is only dust, and to dust he must return:—"*Vous estes poudre, & vous retournéres en poudre*".

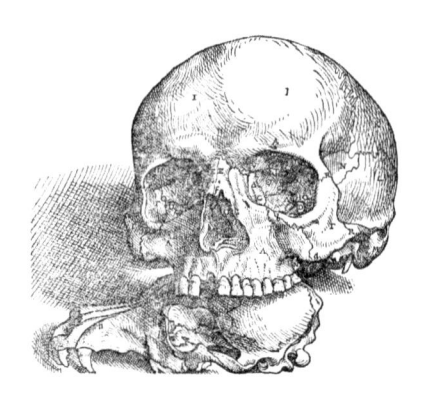

A HUMAN SKULL RESTING ON THE SKULL OF A DOG
(From the "Fabrica", 1543)

CHAPTER TENTH
Publication of the Epitome

Upon the thirteenth day of August, 1542, Vesalius finished the *Epitome* of his great book. The text and illustrations for it were forwarded to Basel by the same merchant who conveyed the manuscript and drawings of the *Fabrica*. The title of the lesser work is as follows:—

> ***Andreae Vesalii Bruxellensis, Scholae medicorum Patavinae professoris, suorum de Humani corporis fabrica librorum Epitome. Basil., et officina Joannis Oporini, Anno, 1543, mense Junio.***

This work is extremely rare. It belonged to the class of *Fliegende Blätter* and was issued unbound. Perfect copies of it are rarely found. The first twelve sheets are printed on both sides; the two last leaves are printed on one side only, in order that they might be cut out and pasted together to show two complete figures. Hence these sheets are often lacking. The *Epitome* appeared in the same year and in the same month as the *Fabrica*, but the latter work was printed first.

The *Epitome* is dedicated to Philip, the son of Charles the Fifth, who, after his father's abdication, was known as Philip the Second of Spain. The title-page is printed from the same plate as the larger work; and Vesalius's portrait also is present. From the fact that the dedication bears the inscription: *Patavii, idibus Augusti 1542*, the erroneous opinion arose that this work preceded the *Fabrica*.

TITLE-PAGE OF VESALIUS'S "EPITOME", 1543

Among the illustrations found in the *Epitome* are seven that are not in the large book; namely, five myologic plates, and the figure of a naked man and one of a woman. The myologic figures in the *Epitome* differ from those in the *Fabrica* in this respect: the muscles are drawn in their natural position, group, and order, so that the surgeon, in treating wounds and in performing operations, may have the correct relations of the parts in mind. Also, the one side of the figure differs from the other: the one showing the superficial muscles, while the other exhibits the deeper musculature. The muscles in the *Fabrica*, with the exception of four complete myologic figures, are represented as they appear in anatomical demonstrations, particular attention being given to their origins and insertions. For the purpose of the artist, the best figures are the three skeletons and the four complete myologic figures which are found in the *Fabrica*.

Two beautiful copies of the *Epitome*, printed on vellum, are in existence. One is in the British Museum and is thought to be the copy which was owned by the celebrated Dr. Richard Mead; the other one is in the possession of the University of Louvain.

Vesalius speaks modestly of the *Epitome*, which he regards as an index or appendix of the *Fabrica*, and is for the use of beginners in anatomy.

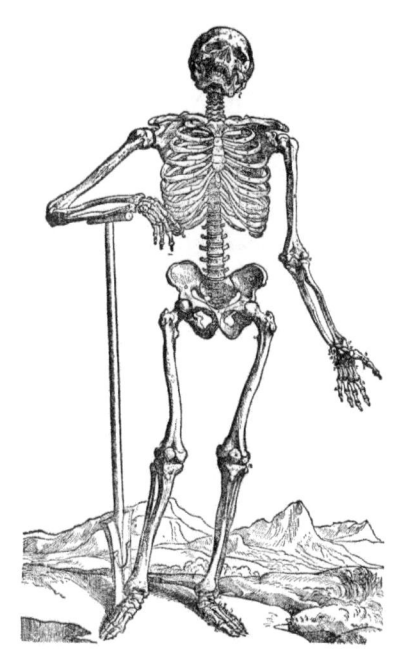

SKELETON BY VESALIUS
(From the "Fabrica", 1543. Reduced one-half)

CHAPTER ELEVENTH
Contents of the Fabrica

The reputation of Vesalius rests securely upon the *Fabrica*. This grand book, which is dedicated to Charles the Fifth, consists of six hundred and fifty-nine folio pages of text; thirty-four pages of index, disposed in three columns to the page; six pages of preface; and two pages of a letter which is addressed to "Joannes Oporinus, the renowned professor of Greek letters in Basel". The work is printed in excellent style. The printed page measures 8 by 12½ inches, including the marginal notes. There are fifty-seven lines to a page, averaging twelve words to a line, or approximately seven hundred words to a page. This was written, amid many duties and distractions, in the short period of three years. It is truly a monument of diligence.

The text of the *Fabrica* is clear and concise; it describes what has to be described and does it well. The errors which Vesalius rectified, and the improvements which he made in anatomy, are so numerous that references can be made to only a few of them. His anatomical writings are of such bulk that they cannot be reviewed adequately within the limits of the present chapter. As regards the *Fabrica*, we may say, with Richardson, that "The dissections and the plates are the book".

FIFTH VESALIAN PLATE OF THE MUSCLES
(From the "Fabrica", 1543. Reduced one-half)

The *Fabrica* contains the rudiments of anthropology as well as the first illustrations of comparative anatomy. Vesalius portrays a human skull resting upon the skull of a dog. He also shows a simian and a canine sacrum and coccyx, to prove his contention that Galen's anatomy was derived from dissection of the lower animals. The *Fabrica* is more than an anatomy. Throughout the work physiology goes hand in hand with the anatomical description. The use and function of each part of the body is given in short, clear sentences.

The *Fabrica* is built upon a practical plan. It treats of anatomy in a logical manner and is composed of seven books, which deal with the following subjects: (1)—Bones and Cartilages; (2)—Ligaments and Muscles; (3)—Veins and Arteries; (4)—Nerves; (5)—Organs of Nutrition and Generation; (6)—Heart and Lungs; and (7)—Brain and Organs of Sense.

The First Book

Vesalius devotes one hundred and sixty-eight pages to the bones and cartilages, treating these structures with a thoroughness that amazed his contemporaries. He was the first author who correctly described the osseous system as a whole. In numerous instances Vesalius places himself in direct opposition to the opinions of Galen. He denied the existence of the intermaxillary bone in adults, and showed that the inferior maxilla does not consist of two pieces, as has been asserted by Galen. The seven bones of the sternum were reduced to three by Vesalius. He denied Galen's statement that the bones of the symphysis pubis separate during parturition. He was the first anatomist to give an accurate description of the sphenoid bone. A small aperture at the root of the pterygoid process of the sphenoid bone is called *foramen Vesalii*. Vesalius proved the existence of marrow in the bones of the hand, which had been denied by Galen. In all respects, he wrote more intelligently of the bones than any anatomist who had preceded him.

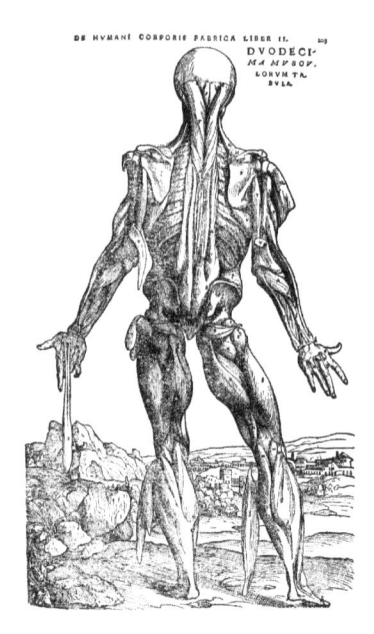

DEEP MUSCLES OF THE BACK BY VESALIUS
(From the "Fabrica", 1543. Reduced one-half)

ANDREAE VESALII
BRVXELLENSIS, DE HVMANI CORPO-
RIS FABRICA LIBER PRIMVS, IIS QVAE
uniuerfum corpus fuſtinent ac fuffulciunt, quibuſq́ omnia
ſtabiliuntur & adnaſcuntur dedicatus.

QVID OS, QVISQVE IPSIVS VSVS
& differentia. Caput I.

S CAETERARVM hominis partium eſt durisſi
mum, & aridiſsimū, maximeq́ terreſtre & frigidum,
& ſenſus denique præter ſolos dentes expers. Huius
enim temperamenti ſummus rerum opifex Deus
ſubſtantiā meritò efformauit, corpori uniuerſo fun
damenti inſtar ſubijciendam. Nam quod parietes &
trabes in domibus, & in tentorijs pali, & in nauibus
carinæ ſimul cum coſtis præſtant, id in hominis fa
brica oſsium præbet ſubſtantia. Oſsium ſiquidem
alia roboris nomine tanquam corporis fulcra pro-
creantur, è quorum numero ſunt tibiarum & femo-
rum oſſa, & dorſi uertebræ, ac omnis ferè oſsium con
textus. Alia reliquis partibus ueluti propugnacula,
tutiſsimiq́ ualli & muri à natura obijciuntur, quem

PART OF THE FIRST TEXT-PAGE OF THE "FABRICA"

The Second Book

Vesalius devotes one hundred and eighty-eight pages to a description of the ligaments and the muscles. This part of his treatise, while it contains a few errors and does not reach the high plane of the first book, is superior to any work of its kind that had preceded it. Vesalius was the first writer to describe the internal pterygoid muscle. He denied the existence of a general muscle of the skin, and stated that the intercostal muscles merely separate the ribs without expanding or contracting the thorax. He held the view that nerves and muscles do not stand in any relation of proportionate strength to one another, large nerves often being distributed to small muscles. He also held that the tendons are similar in structure to the ligaments.

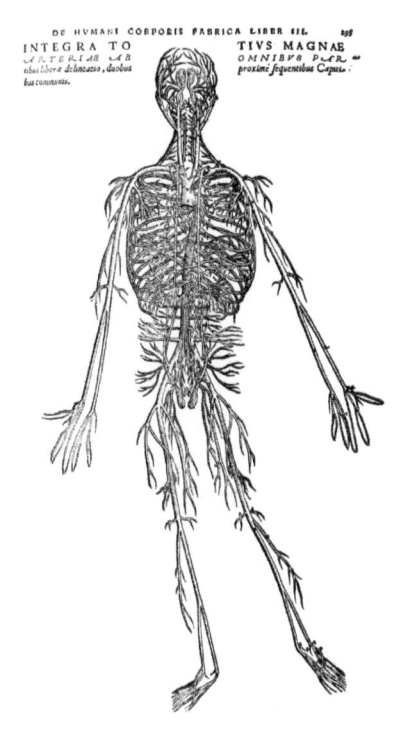

PLATE OF THE ARTERIAL TREE BY VESALIUS
(From the "Fabrica", 1543. Reduced one-half)

Vesalius's plates of the superficial muscles are among the most beautiful that have ever appeared. They have been copied in practically all later treatises on anatomy, and have been used extensively by art-anatomists. His plates of the deeper muscles, while naturally not so pleasing to the eye, are wonderfully near accuracy. The different muscles are drawn to show function as well as structure.

The Third Book

The third part of the *Fabrica*, comprising sixty pages, is devoted to the veins and arteries. Vesalius begins with the definition of a vein, and describes the structure of these vessels in general. The term "artery" is treated in like manner. He introduces several small illustrations which serve to elucidate this part of the text. His first large plate in this section is devoted to the venae portae. This is followed by a full-page picture of the entire venous system. The arterial system is fully described and elaborately illustrated. To these is added another plate, in which both arteries and veins are represented in their natural order. In other plates he shows the special circulations—cerebral, portal, and pulmonary.

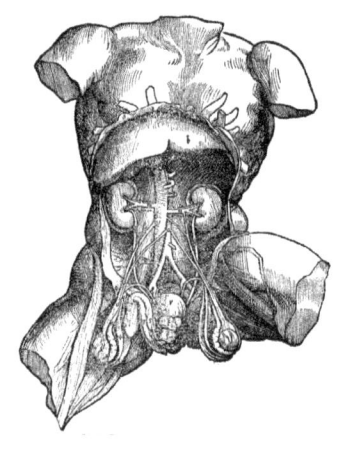

DISSECTION OF THE ABDOMEN BY VESALIUS
(From the "Fabrica", 1543)

Vesalius described the valve which guards the foramen ovale in the foetus, and also noticed the valve-like fold which guards the entrance of each hepatic vein into the inferior vena cava. He also gave an admirable description of the vena azygos. Blinded by the ancient theory of the movement of the blood—a sort of flux and reflux in the veins, he overlooked the function of the venous valves. He described them as eminences, or projections, or accidental rugosities, which in no way interfere with the flux and reflux of the blood.

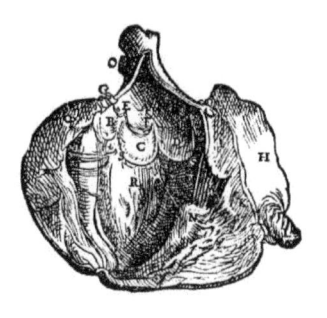

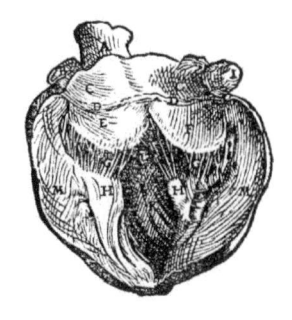

DISSECTION OF THE HEART BY VESALIUS
(From the "Fabrica", 1543)

The Fourth Book

Vesalius devotes forty pages to the cerebral and spinal nerves. The anatomy of the brain is treated in the seventh book. His representations of the nerves are very creditable. He mentions eleven pairs of cranial nerves: the olfactory, the optic, the motores oculorum, the trifacial, the abducens, the portio dura, the portio mollis, the glosso-pharyngeal, the pneumogastric, and the spinal accessory.

His account of the brain—contained in the seventh book—is elaborately minute considering the time when it was written. His illustrations and description of this organ surpass those of scores of later authors. Vesalius fully describes the position of the brain; the membranes which cover it; the cavities, or ventricles, within it; the divisions of cerebrum, cerebellum, and medulla; the anatomy of the base, and the origins of the cerebral nerves. These structures are illustrated from different points of view.

The Fifth Book

The fifth book, comprising more than one hundred pages, is devoted to the organs of nutrition. Here we find an admirable account of the peritoneum, the mesentery, the omentum, the stomach and intestines, the liver, the spleen, and the genito-urinary tract—all of which structures are described and fully illustrated. In this book Vesalius also describes the foetus in utero.

The Sixth Book

In less than fifty pages Vesalius describes the contents of the thorax. He writes intelligently of the membrane lining the thorax, and then gives an account of the arteria aspera, as the trachea was formerly named. Passing on to the lungs, he next takes up the anatomy of the heart. He describes its position, form, and structure in better terms than had been done by preceding anatomists. The auricles, ventricles, and valves are carefully examined. His illustrations of both lungs and heart are excellent.

In the 1543 edition of the *Fabrica*, Vesalius adopts the erroneous view of Galen that openings exist in the septum of the heart. In the second edition of his book, published in 1555, he says that influenced by the views of Galen, he believed that the blood passes from the right to the left ventricle of the heart, through the septum, by means of the pores. Vesalius immediately adds that the septum of the heart is as dense and compact as the rest of this organ, and that not the smallest quantity of blood passes through the septum.

His account of this subject is best given in his own words:—"In recounting as above the structure of the heart, and the use of its different parts, I have followed in the main the doctrines of Galen; not that I regard them in all particulars as consonant with the truth, but because, in attributing new functions and uses to a number of parts, I am still distrustful of myself, and not long ago should hardly have ventured to differ from that Prince of Physicians by so much as a finger's breadth. As for the dividing wall, or septum, between the ventricles forming the right side of the left cavity, the student of anatomy should consider carefully that it is equally thick, compact, and dense, with all the rest of the cardiac substance enclosing the left ventricle. And accordingly, notwithstanding what I have said about the pits in this situation, and at the same time not forgetting the absorption by the portal vein from the stomach and intestines, I still do not see how even the smallest quantity of blood can be transfused, through the substance of the septum, from the right ventricle to the left".

Vesalius and other anatomists knew of the hepatic circulation, or at least believed in some communication between the portal and hepatic veins:—"The branches of this vein"—vena cava—"distributed through the body of the liver, come in contact with those of the portal vein; and the extreme ramifications of these veins inosculate with each other, and in many places appear to unite and be continuous".

Vesalius knew that in several particulars the accepted physiology of the vascular system was wrong. If he could have lived a few years longer, it is possible that he might have solved the great problem which was made clear by William Harvey. In the light of our present knowledge some of Vesalius's words are suggestive:

"When these matters are taken into account, many things at once present themselves in regard to the arterial system, which deserve careful consideration; especially the fact that there is hardly a single vein going to the stomach, the intestines, or even the spleen, without its accompanying artery, and that nearly every member of the portal system has a companion artery associated with it in its course. Again, the arteries going to the kidneys are of such size that they can by no means be affirmed to serve merely for regulating the heat of these organs; and still less can we assert that so many arteries are distributed to the stomach, intestines and spleen for that purpose alone. And there is, furthermore, the fact, which we must for many reasons admit, that there is through the arteries and veins a mutual flux and reflux of materials, and that within these vessels the weight and gravitation of their contents has no effect".

The Seventh Book.

In the seventh book, consisting of less than sixty pages, Vesalius fully describes the anatomy of the brain, of the cranial nerves, and of the organs of sense. His description of the eye is not as near accuracy as might be expected. He places the crystalline lens in the centre of the globe. His description of the organ of vision was only slightly better than that which was given by Galen. Vesalius showed, however, that the optic nerve is not a hollow tube, and that it does not enter the eyeball exactly in the antero-posterior axis.

Conclusion

Considering the time in which he lived, Vesalius was remarkably free from errors. Although to him the arteries were carriers of vital spirits, the veins were the true blood vessels, and, according to the first edition of his great book, the septum of the heart was filled with foramina; yet, we must say with Baas, "these are all mere shadows necessary to the brilliancy of the picture".

Vesalius was more than an anatomist. As a practical physician he had the highest reputation among his contemporaries. He was an accomplished scholar and was thoroughly conversant with the weaknesses of human nature, as is evident from many satirical touches in his writings. Although his great work contains many errors that a tyro of the present day would laugh at, it laid the foundations of our knowledge. Vesalius overthrew the idol of authority in anatomy and taught us to look at Nature with our own eyes.

Portal[24] has paid a splendid tribute to Vesalius. "Vesalius", he says, "appears to me one of the greatest men who ever existed. Let the astronomers vaunt their Copernicus, the natural philosophers their Galileo and Torricelli, the mathematicians their Pascal, the geographers their Columbus, I shall always place Vesalius above all their heroes. The first study of man is man. Vesalius has this noble object in view, and has admirably attained it; he has made on himself and his fellows such discoveries as Columbus could make only by travelling to the extremity of the world. The discoveries of Vesalius are of direct importance to man; by acquiring fresh knowledge of his own structure, man seems to enlarge his existence; while discoveries in geography or astronomy affect him but in a very indirect manner".

Like Harvey, Vesalius was obliged to defend his writings from fierce attacks. The most desperate of his opponents was his old master, Jacobus Sylvius, who was so wedded to the Galenic teachings that he asserted that since Galen's time the thigh bones had changed their shape. He spoke of Vesalius as a "madman, Vesanus, whose pestilential breath poisons Europe". Ponderous discussions were carried on between the friends and opponents of the great anatomist. The complete overthrow of the Galenists resulted.

If Vesalius had remained professor of anatomy in Padua, instead of being appointed physician to Charles the Fifth, at Madrid, in 1544, it is probable that the circulation of the blood would have been discovered by him.

In recent years attempts have been made to show that it was not Vesalius, but Leonardo da Vinci, who was the founder of modern anatomy. A considerable amount of controversial literature has accumulated on this subject. For our purpose it may suffice to quote the conclusions of McMurrich[25]:—"Leonardo was the first to create a new anatomy, but he created it for himself alone; Vesalius demonstrated a new anatomy to the world. It was the publication of Vesalius's *Fabrica* that revolutionized anatomy, while Leonardo's drawings were lying unpublished, at first the cherished possessions of his favorite pupil Melzi, later in the Ambrosian Library in Milan, and still later forgotten in the Royal Library at Windsor. We must credit Leonardo as being the forerunner of the new anatomy, but Vesalius must be recognized as its founder".

INITIAL LETTER BY VESALIUS
(From the "Fabrica", 1543)

CHAPTER TWELFTH
Contemporary Anatomists

Shortly after the publication of the *Fabrica*, great activity was manifested in anatomic research, and numerous opponents and critics of Vesalius appeared in the arena of science. The criticism of such men as Jacobus Sylvius and John Dryander, while it was of a violent type, was of much less importance than was that of Eustachius, Columbus and Fallopius. Vesalius was not without his partisans, of whom Ingrassias and Cannanus are worthy of mention.

Bartholomeus Eustachius

Eustachius was born at San Severino, a small city near Salernum, about the year 1520. He studied anatomy in Rome and made remarkable progress in this science. In the year 1562, as he informs us in his *Opuscula Anatomica*, he was professor of medicine in the Collegio della Sapienza at Rome. Like many other men of genius, Eustachius died in poverty. In August, 1574, having been called by the illness of Cardinal Rovere to Fossombrone, Eustachius died upon the journey.

To Eustachius posterity is indebted for a series of splendid copperplate engravings which were designed to illustrate the anatomy of the human body. These plates, the handiwork of Eustachius, and the first anatomical illustrations wrought in copper, were completed in 1552, only nine years after the first impression of the book of Vesalius. Unfortunately for himself, and worse for medical science, Eustachius was unable to publish them. If this magnificent atlas of anatomy could have been published when completed, the anatomical discoveries of the eighteenth century would have come two hundred years earlier. Unfortunately the entire text of the work is lost. For one hundred and thirty-eight years the Eustachian plates remained either in the family of Pinus, an intimate friend of the anatomist, or were buried in the Papal Library at Rome. When discovered they were presented by Pope Clement XI. to his physician, Lancisi, who published them with notes of his own, at Rome, in 1714. In 1740 they were issued under the direction of Cajetan Petrioli. Four years later the edition by Albinus appeared, which was republished in 1761. The anatomical writings of Eustachius were published during his lifetime, in 1564. It is upon his *Tabulae Anatomicae* that the fame of this wonderful man is founded. If this work had been published in 1552, Eustachius would have divided with Vesalius the honor of founding human anatomy. The victim of circumstances, his name has been overshadowed by that of Vesalius, to whom in some respects he was superior. Deprived during life of his merited honors, Eustachius has been awarded a goodly share of posthumous fame.

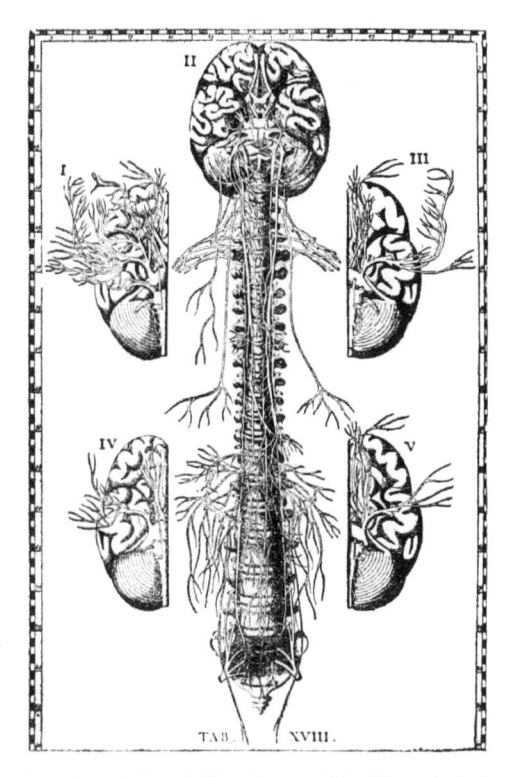

BRAIN AND NERVES BY EUSTACHIUS
(Reduced one-half)

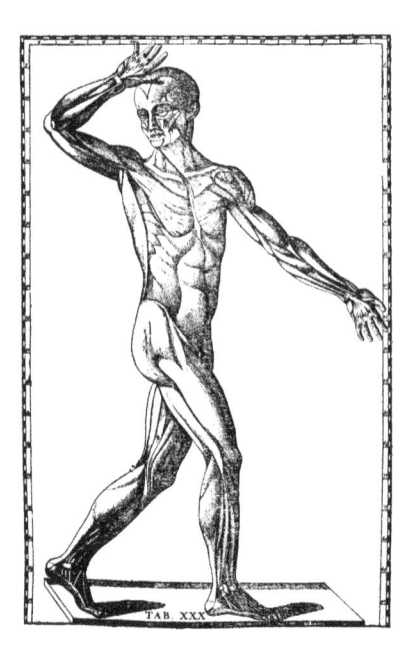

MUSCLES BY EUSTACHIUS
(Reduced one-half)

Eustachius was the first anatomist to describe, with any degree of accuracy, the tube which bears his name. We can truly say he discovered it, since Alcmaeon dissected only the lower animals, and was not an accurate observer, as his view that goats breathe through the ears, amply testifies. Eustachius discovered the tensor tympani and stapedius muscles, the modiolus and membranous cochlea, and the stapes. The honor of the discovery of the stapes is claimed for no less than five renowned anatomists, namely, Fallopius, Ingrassias, Columbus, Colladus, and Eustachius. It is unnecessary to discuss this disputed claim to priority. The truth seems to be that the stapes was discovered by both Ingrassias and Eustachius, each independently of the other. In 1546 Ingrassias publicly demonstrated the little bone of the ear in his lectures at Naples. Fallopius, after learning from an eyewitness that Ingrassias had actually discovered and named the ossicle, relinquished his claim to the discovery. Columbus and Colladus filed their information at too late a date. Eustachius, as previously stated, finished his anatomical plates in 1552. His seventh plate shows, among other subjects, the auditory ossicles—malleus, incus and stapes—and tensor tympani muscle. These objects are delineated as taken from a human subject, and also from a dog.

Eustachius discovered the origin of the optic nerves, and the sixth cerebral nerves. He gives excellent pictures of the corpora olivaria and corpora pyramidalia; of the stylo-hyoid muscle; of the deep muscles of the neck and throat; of the suprarenal capsules, and of the thoracic duct. He also described the ciliary muscle. Eustachius was the first anatomist who accurately studied the teeth and the phenomena of the first and second dentition. In his researches he employed magnifying glasses, maceration, exsiccation, and various methods of injection.

Realdus Columbus

The first anatomical treatise containing an account of the lesser, or pulmonary circulation, was the monumental work, *De Re Anatomica, libri xv.*, written by Realdus Columbus and sumptuously published at Venice in the year 1559. This, however, was not the first printed account of the lesser circulation. Six years prior to the publication of the book of Columbus, the unfortunate Servetus, in a theological treatise, described correctly the course of the blood in its transit through the lungs. Tried for heresy, Servetus was burned, together with all obtainable copies of his book. Although it had been printed, the work was suppressed; hence it follows that Columbus was the first to publish the great discovery. Of the life of this anatomist we know but little. Born at Cremona, a small Milanese village, the year of his birth is unknown. He died in 1559, while his book was being printed. A few copies were finished before his demise, since a copy belonging to the late Dr. George Jackson Fisher, of Sing Sing, N.Y., contains the author's own dedication to Pope Paul IV., while in other exemplars, the dedication has been written by the two sons of Columbus, and is addressed to "*Pio IIII., Pont. Max*". This prelate, on the death of Paul IV., on August 18, 1559, became the head of the Church.

Some writers have held that the discovery of the lesser circulation was not made by Columbus independently of Servetus, but that a copy of the book of Servetus had drifted into Italy and had been read by Columbus. There is no direct evidence to support this view. When Vesalius was called to Madrid as physician to Charles the Fifth, Columbus, in 1544, succeeded him in the University of Padua; two years later he filled the anatomical chair at Pisa, and in 1546, Pope Paul IV. called him to Rome. Here he spent the later years of his life, engaged in teaching anatomy and in writing his book. For forty years Columbus pursued his anatomical studies, and in that period he dissected an unusually large number of bodies. Fourteen subjects passed under his scalpel in a single year.

TITLE-PAGE OF COLUMBUS'S ANATOMY
(Reduced one-half)

Columbus frequently made experiments upon living animals. He was the first to use dogs for such purposes, preferring them to swine. Book XIIII. of the work of Columbus is upon the subject of vivisection, *De viva sectione*. In this he tells us how to employ living dogs in demonstrating the movements of the heart and brain, the action of the lungs, etc. Columbus was the first anatomist who demonstrated experimentally that the blood passes from the lungs into the pulmonary veins. "When the heart dilates", says Columbus, "it draws natural blood from the vena cava into the right ventricle, and prepared blood from the pulmonary vein into the left; the valves being so disposed that they collapse and permit its ingress; but when the heart contracts, they become tense, and close the apertures, so that nothing can return by the way it came. The valves of the aorta and pulmonary artery opening, on the contrary, at the same moment, give passage to the spirituous blood for distribution to the body at large, and to the natural blood for transference to the lungs".

Like Servetus, Columbus held to the idea of "spiritus". Harvey was the first physiologist who recognized the circulation as purely a movement of blood. All before him assumed the existence of a mixture of air and blood. Columbus, pupil and prosector of Vesalius, like his great master, denied the existence of foramina in the cardiac septum.

Gabriel Fallopius

GABRIEL FALLOPIUS

Gabriel Fallopius (1523-1562), of Modena, was a noted Italian anatomist. In his twenty-fifth year he was made professor of anatomy at Pisa. Although the span of his life was short, he will be remembered always as the discoverer of the tubes which bear his name. According to Fisher, Fallopius "described the ear more minutely than had ever before been done. He discovered the little canal along which the facial nerve passes after leaving the auditory; it is still called the *aquaeductus Fallopii*. He demonstrated the fact of the communication of the mastoid cells with the cavity of the tympanum; and also described the fenestrae rotunda and ovalis. In the treatment of diseases of the ear, he used an aural speculum, and employed sulphuric acid for the removal of polypi from the meatus. In some of his supposed discoveries he had long been anticipated; for example, the tubes which bear his name were known and accurately described by Herophilus, over three hundred years before the Christian era, and also by Rufus of Ephesus, of whom Galen speaks as the best anatomist of the second century. Rufus refers to two varicose and tortuous vessels passing from the testes (as the ovaries were called) to the cavity of the uterus. Fallopius, however, gave a full account of their course, position, size and structure. He cut into them and found them hollow, gave them the name of tubae seminales, and posterity attached his name to them, and in time came to a better comprehension of their true function. This is not the only instance in the history of anatomical discovery where the name of a person, not its discoverer, has been given to an organ. Allusion has been made to Fallopius as a botanist; a genus of plants, *Fallopia*, has been named in honor of him".

Fallopius was appointed professor of anatomy at Pisa, in the year 1548; and later, at the instance of the Grand Duke of Tuscany, Cosimo I., he received a professorship at Padua, as successor to Vesalius. Besides the chair of anatomy and surgery and of botany, he also held the office of superintendent of the new botanic garden in that city. Fallopius remained in Padua to the day of his death, which occurred in 1562. He was very properly succeeded by his favorite pupil, Fabricius ab Aquapendente, who had been for some time previously his anatomical demonstrator. His collected works, as published in Venice, 1606, embrace twenty-four treatises distributed in three folio volumes. Only one of his works was published during his lifetime, namely, his *Observationes Anatomicae*, Venice, 1561, which is considered one of his most valuable books, containing, as it does, most of his discoveries and his animadversions on the works of other anatomists.

This was written as a supplement to the anatomy of Vesalius, for it follows the same order, passes upon the same subjects, corrects the inaccuracies of the Vesalian treatise, and supplies what is wanting. Throughout the work Fallopius treats Vesalius with great respect, and never mentions him without an honorable title. Vesalius wrote an answer to this work, entitled, *Observationum Fallopii examen*, in which he acknowledges the courtesy of

Fallopius, but, as argument progresses, appears to be out of temper.

After the death of Fallopius it was thought that no successor except Vesalius could be found competent to fill his place. Accordingly Vesalius was chosen. The news of his appointment reached him while he was returning from a pilgrimage to Jerusalem. Unfortunately he was shipwrecked and perished, otherwise history would have afforded an example of the master filling the chair of the pupil.

John Philip Ingrassias

Ingrassias, who lived between the years 1510-1580, was a graduate of the celebrated Paduan School. He described minutely the anatomy of the ear, including the tympanum, fenestrae rotunda and ovalis, the cochlea, the semi-circular canals, and the tensor tympani muscle. His admiring pupils caused his portrait to be painted and placed in the Neapolitan School, with this inscription:—"To Philip Ingrassias, of Sicily, who, by his lectures, restored the science of true Medicine and Anatomy in Naples, his pupils have suspended this portrait as a mark of grateful remembrance". Ingrassias was a voluminous writer, his chief work being a treatise on osteology, which was published twenty-three years after his death. When the plague depopulated Palermo, in 1575, his devotion was such as to earn for him the title of the Sicilian Hippocrates. Few men have been more earnest workers in medical science. If his fame as an anatomist has not equalled that of others, the cause is to be sought in the multiplicity of competitors, not in lack of zeal and ability.

INGRASSIAS

CHAPTER THIRTEENTH
Commentators and Plagiarists

Medical history furnishes numerous examples of literary theft. In many instances an entire set of anatomical plates has been pirated by unscrupulous publishers. In a few cases both text and plates have been appropriated by medical authors. The most notorious example of this form of theft was furnished by William Cowper (1666-1709), an English surgeon and anatomist, who, having secured three hundred copies of Bidloo's set of one hundred and five anatomical plates, in 1697 issued the work[26] as his own. Cowper added a few original illustrations to the book.

Vesalius suffered severely at the hands of the plagiarists. Pirated editions of the *Tabulae Anatomicae* were printed in several cities, chiefly in Germany. As regards the *Fabrica*, we may say that it has been the fountain from which many anatomical writers have derived practically all of their illustrations and much of their text.

The fame of the *Fabrica* soon spread throughout Europe. It was published in Germany, in Holland and in England. An epitome of its contents was issued in Latin, in 1545, by Thomas Geminus, or Gemini, under the title:—*Compendiosa totius Anatomiae delineatio, aere exaratum per Thomam. Geminum.* It contained forty of the Vesalian plates, cut in copper, and was the first book issued in England in which the roller printing process was employed. It was dedicated to Henry the Eighth, and was embellished with "one of the earliest and most curious of all extant engraved title-pages".

In 1553, Geminus issued a second edition, in which the text was translated into English. This edition was dedicated to Edward the Sixth, with a commendatory note, "To the gentill readers and Surgeons of Englande". Six years later the third English edition appeared, which was inscribed to Queen Elizabeth. It contains the first published portrait of the Queen. She is shown upon the engraved title-page, and, strange to say, above her is another queenly figure, with a pen in her right hand, a wreath on her left, her foot resting on the globe, and styled *Victoria*.

Another English work on anatomy, which is filled with poor imitations of Vesalius's illustrations, is the *Microcosmographia* of Helkiah Crooke, or Crocus, who was "Professor in Anatomy and Chirurgery". Its chief value rests in an elaborately engraved title-page, a part of which shows Crooke giving a demonstration in anatomy in the presence of the "Worshipfull Company of Barber-Chirurgeons", in London, early in the seventeenth century.

John Banister of Nottingham, in 1578, borrowed a few Vesalian woodcuts for use in *The Historie of Man, sucked from the sappe of the most approved Anatomists and published for the Utilitie of all Godly Chirurgians within this Realme.*

Most of the host of translators, epitomizers, commentators and imitators of Vesalius have passed into oblivion. A few of these persons have possessed enough of individuality to deserve recognition.

Juan Valverde di Hamusco, a Spaniard who was born about the year 1500, studied anatomy at Padua and later at Rome. His book, *Historia de la Composicion del Cuerpo Humano*, was published at Rome in 1556. It contains forty-two copperplates and an engraved title-page. Although the author says he has used only the Vesalian plates, his work contains several plates which are not to be found in Vesalius's writings. For example, Valverde shows a *muskelmann* with his skin held in his right hand, the left grasping a dagger which may have been used in the skinning process. Other original drawings show the abdomen and intestines, a pregnant woman with the abdomen opened, and illustrations of the superficial veins.

Valverde was physician to Cardinal Juan de Toledo, Archbishop of Santiago, to whom the work is dedicated. The illustrations were drawn by Gaspar Becerra and were engraved by Nicholas Beatrizet. Valverde's book went through several editions. It forms a landmark in the medical history of Spain—a country which, for many years, was behind other states of Europe in matters of science.

To name the list of anatomical writers who have derived their artistic inspiration from the *Fabrica* would require much more space than is at our disposal. It must suffice to say, that, for a period of two centuries, nearly all treatises on anatomy contained illustrations which were taken from the writings of Vesalius. With few exceptions, these reproductions were little better than caricatures of the original figures.

Of the numerous editions of the *Fabrica* there are three which are highly prized, namely, the first one, 1543; the second, issued in 1555, containing eight hundred and twenty-four pages, with many changes in the text; and the 1725 edition of the collected writings of Vesalius. The last named is a huge volume which was published at Leyden under the supervision of Boerhaave and Albinus, with the illustrations cut in copper by Jan Wandelaar[27].

It contains the *Fabrica*, the *Epitome*, the *Epistola de Radicis Chynae*, various anatomical treatises of a controversial character, and the *Chirurgia Magna* which has been wrongly attributed to Vesalius. Morley says of this book:—"After his death a great work on surgery appeared, in seven books, signed with his name, and commonly included among his writings. There is reason, however, to believe that his name was stolen to give value to the book, which was compiled and published by a Venetian, Prosper Bogarucci, a literary crow, who fed himself upon the dead man's reputation".

CHAPTER FOURTEENTH
The Court Physician

Vesalius, having finished the *Fabrica*, intended to write a work on the practice of medicine which should be based on pathology. He makes mention of this in the preface of the *Fabrica*, and in numerous places in the body of the book he describes the pathologic appearances which he found in dissection.

Returning to Padua after a year's absence, he found that the University for which he had strenuously labored was a very hotbed of opposition. His former pupil and friend, Realdus Columbus, who was now lecturing on anatomy at Padua, had turned against him. How deeply Vesalius was wounded by the man whom he had made, can be appreciated only by those who have been placed in similar circumstances. The controversy between Columbus and Vesalius was of a bitter and personal character.

On all sides the views of Vesalius were attacked, and the defenders of Galen joined hands with men like Columbus in an effort to besmirch the great anatomist. Disgusted with such treatment, Vesalius, early in 1544, went to Pisa. Here he conducted a course in anatomy. Leaving Pisa, he went to Bologna where he made some special dissections upon two bodies. About this time he declined a chair in the University of Pisa which was tendered to him by direction of Cosimo de' Medici. Tired of the apparently useless effort to make men see the truth, sick of disputes and arguments, persecuted by members of his own profession, in a fit of passion Vesalius threw his manuscripts into the fire and ended his career as a scientist. "Thus", says Morley, "he destroyed a huge volume of annotations upon Galen; a whole book of Medical Formulae; many original notes upon drugs; the copy of Galen from which he lectured, covered with marginal notes of new observations that had occurred to him while demonstrating; and the paraphrase of the books of Rhazes, in which the knowledge of the Arabians was collated with that of the Greeks and others".

CHARLES THE FIFTH

While in this frame of mind it is not surprising that he should have accepted the appointment of Archiatrus to Charles the Fifth of Spain.

The great Emperor was now at the zenith of his fame. His kingdom, which reached from South America to the Zuyder Zee, was well under control, but the monarch already contemplated the abdication of the throne in favor of his son Philip, who is known in history as Philip the Second.

Vesalius left Italy and took up his residence at Madrid. He was now in his thirtieth year. As Archiatrus he accompanied the Emperor in the fourth French war, in which he gained his first experience as a military surgeon. He also acted as physician to Charles and to the members of the imperial household. The war ended in September 1544. In January, 1545, Charles went to Brussels, and remained in the Netherlands for many months. Vesalius was now in his native country, and in April, 1546, he visited the graves of his ancestors at Nymwegen and Wesel. In the same year he published a new edition of his treatise on the China root.

On the twenty-fifth day of October, 1555, amid a scene of pomp and splendor, in the presence of the assembled representatives of the Netherlands, Charles formally surrendered to his son all his territories, jurisdiction and authority in the Low-Countries. This was the first of a series of acts by which the Emperor gradually relinquished the reins of power, in order to spend his remaining days in a cloister. Philip thus became the heir to a vast dominion. Vesalius was continued in office as Archiatrus by the new Emperor. From both Charles and Philip, Vesalius received many marks of honor. It was he who rescued Charles from what was thought to be a mortal disease. At a later date, when Philip's unfortunate son, Don Carlos, received a severe injury to the head, and after the treatment of the Spanish physicians had failed, it was Vesalius who saved his life by an operation. These cures, and the accurate prediction of the death-day of Maximilian d'Egmont, placed the fame of Vesalius at high tide.

CHAPTER FIFTEENTH
Pilgrimage and Death

Suddenly, early in the year 1564, for a reason which has never been explained satisfactorily, Vesalius left Madrid. Apparently he was at the height of success. He was famous as a physician and surgeon; he was a favorite at the Spanish court; he had amassed a fortune; and seemingly he was destined to pass his remaining days under the most favorable surroundings. As occurs to all great men, he had excited the jealous animosity of many of the members of his profession. The efforts of the Madrid physicians to ignore the talents of one whom they regarded as a foreigner, long since had reacted to the advantage of the Archiatrus.

PHILIP THE SECOND

During the twenty years that he had filled the post of Archiatrus, the scalpel of Vesalius was rusting: but the controversy concerning the infallibility of Galen was still raging. The violent criticisms of Sylvius upon the *Fabrica* had been silenced by death, but others took up the cause of Galen where Sylvius had left it. But the passing years had brought a new coterie of professors, who, like Fallopius at Padua; Rondelet at Montpellier; Massa at Venice; and Fuchs at Tübingen, were boldly teaching many things that were contrary to Galen.

Life at the Spanish court was not favorable to the study of science. "The hand of the Church", says Foster[28], "was heavy on the land; the dagger of the Inquisition was stabbing at all mental life, and its torch was a sterilizing flame sweeping over all intellectual activity. The pursuit of natural knowledge had become a crime, and to search with the scalpel into the secrets of the body of man was accounted sacrilege. It was for a life in priest-ridden, ignorant, superstitious Madrid that Vesalius had forsaken the freedom of the Venetian Republic and the bright academic circles of Padua; in Madrid, where, as he himself has said, 'he could not lay his hand on so much as a dried skull, much less have the chance of making a dissection'. Moreover, he must have felt the loss of Charles, who, whatever his faults, recognized the worth of intellectual efforts, and in many ways had shown his sympathy with Vesalius's love of knowledge. Such sympathy could not be looked for in the narrow and bigoted Philip".

About this time Vesalius received a copy of the *Observationes Anatomicae* of his pupil Fallopius, who, having learned all that his master had taught of anatomy, continued his studies with great skill and industry. Such a book, coming at an opportune time, must have seemed like a voice calling the Archiatrus back to the intellectual life, bringing to his mind's eye the recollection of his happy days in Italy.

Vesalius travelled to Venice by way of Perpignan. While in Venice he visited the printer, Francesco Sanese, and discussed the publication of a new book which should contain his reply to Fallopius. In a short time he started for Cyprus in company with Jacobo Malatesta, the commander of the Venetian forces in that island. Thence he passed to Jerusalem on a pilgrimage to the Holy Land. Vesalius never returned from that journey. Information of his death reached Brussels towards the end of that year—1564.

What was the reason for this pilgrimage? Various alleged authorities have given different versions, many of which are evidently fictitious. The most reasonable account, which emanates from Spanish-French sources, dates from a letter written January 1, 1565, to the physician Caspar Peucer by Hubert Languer, or Hubertus Languetus, the Huguenot friend of Philip Sidney, which says:—"They say that Vesalius is dead. Doubtless you have heard that he went to Jerusalem. That journey had, as they tell us from Spain, an odd reason. Vesalius, believing a young Spanish nobleman whom he had attended to be dead, obtained leave of the parents to open the body for the sake of inquiring into the cause of the illness, which he had not rightly comprehended. This was granted; but he had no sooner made an incision into the body than he perceived the symptoms of life, and opening the breast, saw the heart beat. The parents coming afterwards to the knowledge of this, were not satisfied with prosecuting him for murder, but accused him to the Inquisition of impiety, in hopes that he would be punished with greater rigor by the judges of that tribunal than by those of the common law. But the King of Spain interposed, and saved him on condition that by way of atoning for the error he should undertake a pilgrimage to the Holy Land".

The pilgrimage was made, the Holy Sepulcher was visited, and the weary wanderer had started for Padua to take the chair which was made vacant by the death of Fallopius. A violent storm swept the Ionian Sea. Vesalius's ship was wrecked upon the island of Zakynthos, where, on the fifteenth day of October, 1564, the Archiatrus died of exhaustion.

Such was the miserable end of Andreas Vesalius of Brussels, a man, who, before he had attained his thirtieth year, had become the greatest anatomist that the world has ever seen.

FOOTNOTES

[1]Théorie de la figure humaine. Paris, 1773.

[2]Moehsen: Verzeichnis einer Sammlung von Bildnissen. Berlin, 1771; page 59.

[3]Bell: Observations on Italy. Edinburgh, 1825; page 257.

[4]Galen: De Anatomicis Adininistrationibus. Lib. II.

[5]Celsus: De Medicina. Lib. I.

[6]Fisher: Claudius Galenus. Annals of Anatomy and Surgery, Vol. IV., page 216.

[7]Saint Basil, in his maturer years, deeply regretted that he had studied classical literature in his youth. Jerome regarded the reading of the writings of antiquity as a terrible crime. Gregory the Great declared a knowledge of grammar even for a layman to be indelicate.—Fort: Medical Economy during the Middle Ages. N. Y., 1883; pages 102, 103.

[8]Meryon: History of Medicine. London, 1861; vol. I, page 479.

[9]Adam; Vitae Germanorum Medicorum. Haidelbergae, 1620: page 224.

[10]Zwinger: Theatrum Vitae Humanae. Basileae, 1571.

[11]Vesalius: Fabrica, 1543, preface.

[12]Sylvius: Ordo et Ordinis Ratio in Legendis Hippocratis et Galeni Libris, 1539.

[13]The Collége Royal de France was founded by Francis the First. This enlightened patron of the sciences and arts recognized the merits of scientific men and rewarded them with his money and his friendship. He established the Collége de France with twelve richly-endowed professorships, one of which was devoted to medicine. The lectures were free to all who desired to attend. The first incumbent of the chair of medicine was Vidus Vidius, Guido Guidi, of Florence, who filled this position from 1542 to 1548. Such success followed his labors that, on his return to Italy, his experience in Paris was the subject of this witticism: *Vidus venit, Vidius vidit, Vidus vicit.*

[14]Northcote: History of Anatomy. London, 1772; page 56.

[15]Portal: Histoire de l'Anatomie et de la Chirurgie. Paris, 1770; vol. I, page 365.

[16]Moreau: Vita Sylvii, in Sylvii Opera Medica. Geneva, 1635.

[17]Vesalius: De radice Chinae epistola, 1546; pages 151, 152.

[18]Archives Curieuses de l'Histoire de France.

[19]Guinterius: Anatomicarum Institutionum, 1539.

[20]Paraphrasis in nonum librum Rhazae medici Arabis clariss. ad Regem Almansorem, de singularum corporis partium affectuum curatione, autore Andrea Wesalio Bruxellensi Medicinae candidato. Lovanii ex officina Rutgeri Resii. mense Februar. 1537.

[21]Radicis Chinae usus, Andrea Vesalio autore. Lugd., 1547; page 278.

[22]Moehsen: Verzeichnis einer Sammlung von Bildnissen. Berlin, 1771; page 82.

[23]Sandrart: Teutsche Academie. Nürnberg, 1685: vol. II., page 243.

[24]Portal: Histoire de l'anatomie et de la chirurgie. Paris, 1770; vol. I., page 399.

[25]McMurrich: Medical Library and Historical Journal, December, 1906.

[26]Cowper: The Anatomy of Human Bodies. Oxford, 1697.

[27]Andreae Vesalii Opera Omnia Anatomica et Chirurgica in duos tomos distributa cura Hermanni Boerhaave et Bernhardi Siegfried Albini. Lugduni Batavorum, 1725.

[28]Foster: Lectures on the History of Physiology. Cambridge, 1901, page 17.